CALIFORNIA NATURAL HISTORY GUIDES

INTRODUCTION TO
THE PLANT LIFE OF
SOUTHERN CALIFORNIA

California Natural History Guides

Phyllis M. Faber and Bruce M. Pavlik, General Editors

Introduction to

THE PLANT LIFE OF SOUTHERN CALIFORNIA

Coast to Foothills

Philip W. Rundel
Robert Gustafson

UNIVERSITY OF CALIFORNIA PRESS
Berkeley   Los Angeles   London

We dedicate this book to the memory of Bob Ornduff
and Dwight Billings, colleagues and mentors,
to whom we owe much.

California Natural History Guide Series No. 85

University of California Press
Berkeley and Los Angeles, California

University of California Press, Ltd.
London, England

© 2005 by The Regents of the University of California

Library of Congress Cataloging-in-Publication Data

Rundel, Philip W. (Philip Wilson).
   Introduction to the Plant Life of Southern California: Coast to Foothills /
Philip W. Rundel and Robert Gustafson.
     p.   cm. — (California natural history guides ; 85)
   Includes bibliographical references and index.
   ISBN 0-520-23616-5 (cloth : alk. paper) — ISBN 978-0-520-24199-2 (pbk. : alk.
paper)
    1. Plant communities—California, Southern.   2. Plant ecology—California,
Southern.   I. Gustafson, Robert, 1939–.   II. Title.   III. Series.

QK149.R86   2005
581.9794'9—dc22   2004014332

Manufactured in China
10 14
10 9 8 7 6 5 4 3

The paper used in this publication meets the minimum requirements of
ANSI/NISO Z39.48–1992 (R 1997) (*Permanence of Paper*). ♾

Cover photograph: Post-fire chaparral succession in the Santa Monica Mountains.
Photograph by Robert Gustafson.

The publisher gratefully acknowledges the generous
contributions to this book provided by

the Moore Family Foundation
Richard & Rhoda Goldman Fund
and
the General Endowment Fund of the
University of California Press Associates.

CONTENTS

Mediterranean-Climate Regions

California represents one of only five small regions of the world that possess a mediterranean climate, which is characterized by mild, wet winters and dry summers. The other locations are in central Chile, the Mediterranean Basin of southern Europe and northern Africa, the Cape Region of South Africa, and Southwestern and South Australia (map 1). These regions with this highly unusual climate account for only a tiny portion of the world's land area and occur only on the western margins of continental landmasses between about 30 and 40 degrees latitude. Subtropical high-pressure centers shield these areas from summer storms. Millennia of evolution in the five mediterranean climate regions have produced a remarkable and globally significant degree of diversity among both plants and animals. All five regions are included in a select group of 25 regions around the world designated as key ecological hot spots because of the size and uniqueness of their biota.

The California floristic province, representing one of these important ecological hot spots, is defined not only by a mediterranean climate but also by unique plant relationships. It covers the great majority of the state, excluding only the southwestern desert regions and the Great Basin to the east of the Sierra Nevada and Cascade Range (map 2). The region extends beyond our state northward into the Klamath Mountains of southwestern Oregon and southward into northwestern Baja California.

Southern California Region

Southern California as defined in this book begins at Point Conception, where the shoreline curves south and eastward, and extends along the coast to the Mexican border. The Santa Inez Mountains of Santa Barbara County and the Transverse

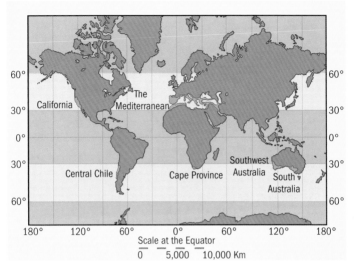

Map 1. Mediterranean-climate regions of the world (map by Lisa Pompelli).

Ranges, including the Tehachapi, San Gabriel, and San Bernardino Mountains, mark the northern boundary of Southern California (pl. 1). The eastern regional boundary extends
southward along the Peninsular Ranges, which include the
San Jacinto and Santa Ana Mountains in Riverside and Orange Counties and the Palomar, Cuyamaca, and Laguna
Mountains to the south in San Diego County. Excluded from
the Southern California region are the central Coast Ranges
and the San Joaquin Valley to the north, and the Mojave and
Sonoran Deserts that lie to the east in the rain shadow of the
Transverse and Peninsular Ranges.

Southern California presents classic mediterranean-climate
conditions. Rainfall is highly seasonal, peaking in winter and
rarely occurring in summer. Periods of six months or more
without rain are not at all unusual. Santa Barbara, near the
northern coastal limit of our region, receives a yearly average of

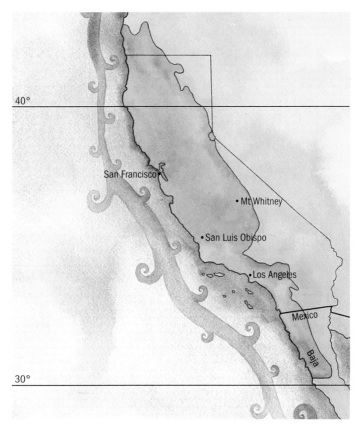

Map 2. The California floristic province (map by Lisa Pompelli).

18 inches of rain. Los Angeles averages only 15 inches and San Diego only 10 inches. Inland, the yearly average is typically 20 to 35 inches in the foothills and increases with elevation on the coastal sides of the mountains. Precipitation, much of it snow, reaches about 45 inches per year in the upper elevations of the San Gabriel, San Bernardino, and San Jacinto Mountains.

Rainfall also varies greatly between years. Los Angeles, for

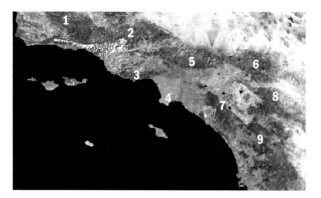

Plate 1. Satellite view of Southern California showing major topographic features: 1. Coast Ranges, 2. Tehachapi Mountains, 3. Santa Monica Mountains, 4. Palos Verdes Peninsula, 5. San Gabriel Mountains, 6. San Bernardino Mountains, 7. Santa Ana Mountains, 8. San Jacinto Mountains, 9. Peninsular Ranges.

example, with an annual average of about 15 inches, has had as little as four inches and as much as 38 inches per year over the past century. The intensity of rainfall over short periods is also remarkable and contributes to landslides and mudflows. The foothills of the San Gabriel Mountains have had up to 25 inches of rain in 24 hours.

The coastal regions virtually never experience freezing temperatures, and thus many plant species with subtropical ancestry survive there. Frost and even snow become regular features as elevations increase. Like rain, however, they can vary greatly from year to year. Hard frosts sometimes extend across the Los Angeles Basin and into desert areas to the east. In such years snowfalls can extend down to the foothill suburbs of Los Angeles. Mount Wilson, at the upper margin of the chaparral zone in the San Gabriel Mountains above Los Angeles, has recorded temperatures as low as nine degrees F.

Over the past few million years a variety of geologic forces have shaped the Southern California landscape. Perhaps the most significant influence has been the ongoing collision of the Pacific Plate with the North American Plate along our coastal margin. Almost all of the Southern California region and most of the coastal ranges of central California as far north as San Francisco lie to the south and west of the San Andreas Fault and thus are part of the Pacific Plate. The majority of California, however, is part of the North American Plate. Contact between these two plates began about 24 million years ago and continues to be a major force shaping our landscape today. As the two plates try to occupy the same physical space, lateral slippage along hundreds of miles of faults has dissipated much of the tremendous force generated, but the remaining pressure has still wrinkled the crust, forcing up mountain ranges and forming deep subsidence basins.

Uplift of the crust and subsequent erosion have been going on for millions of years and continue today. Much of the Los Angeles Basin and the Oxnard Plain of Ventura County, for example, were once deep ocean basins, but they have slowly filled with erosional debris from mountains over the past few million years. Young mountain ranges such as the San Gabriel and San Bernardino Mountains erode even as they continue to be uplifted. Our shoreline, as well, is highly dynamic, with uplift and erosion occurring at the same time. Coastal areas of Ventura County, for example, have been uplifted at a remarkable average rate of six inches per decade for the past 200,000 years, one of the highest rates known in the world.

It is not surprising, therefore, that the Southern California region includes a wide diversity of topographic features. Moreover, its dynamic geologic history has exposed a variety of types of rocks, which have in turn produced a variety of soil conditions. Granites, sedimentary sandstones and shales, volcanic basalts, occasional limestones, and nutrient-deficient metamorphic rock all influence soil types in Southern California. These varied soil conditions, combined with high topographic diversity and the associated climatic gradients, which have experienced dynamic changes over the past 10,000 years, have been ideal for the evolution of varied plant species. Elevation, slope exposure, proximity to the ocean, soil depth, and soil nutrition have profound influences on the distribution and occurrence of plant species and communities.

Plant Diversity

Whether we measure biodiversity by numbers of plant and animal species present or by conservation significance (for example, by numbers of designated rare and endangered species), coastal Southern California rates higher than any other part of California or the continental United States. Here a mediterranean climate combines with diverse topography and dynamic fire cycles to produce a mosaic of habitat types, including chaparral, woodlands and savannas, coastal sage scrub, grasslands, riparian woodlands, wetlands, and coastal marshes (pl. 2).

Plate 2. A mosaic of chaparral and woodlands, the characteristic landscape of Southern California.

The richness and evolution of plant diversity in Southern California depend not only on physical geography and climate, but also on natural disturbances. Fire of varying frequency, intensity, and extent, as discussed in more detail in chapter 5, has been a natural component of the Southern California environment since long before humans first arrived in our region 10,000 to 12,000 years ago, and it still is today, although in a somewhat altered manner (pl. 3). Added to this are irregular extreme climate conditions, which are more significant than ordinary conditions in driving the evolution of plants and animals.

The Southern California region includes approximately 2,200 species of native vascular plants (pl. 4), which constitute nearly half of the flora occurring anywhere within our state. Another 700 species of nonnative plants now grow and reproduce on their own here. The most widespread natural ecosystems of our region are the coastal and interior sage scrub, chaparral, and oak woodlands that cover the great majority of Southern California below the coniferous forests of the higher mountains. Also important but more restricted in size and location are coastal dunes and bluffs, low-elevation conifer woodlands, riparian woodlands, and a variety of wetland habitats such as salt marshes, freshwater marshes, and vernal pools. The plant communities of each are described in the following chapters, beginning on the coast and moving inland. A separate chapter covers the Channel Islands and their unique flora, followed by final chapters about invasive nonnative plant species and issues in preserving biodiversity.

Space limitations allow us to discuss here only about 300 plant species—less than 10 percent of the total number possible. Because our approach is ecological, we have chosen to focus on the most characteristic or most frequently encountered plants. Our selection is further bi-

Plate 3. Fire, a natural factor in Southern California.

Plate 4. Diverse species of spring annuals carpeting the foothills of Southern California.

ased toward woody and semiwoody plants in the most common habitats because these species can generally be recognized at any time of year. We have given particular attention to commonly encountered species in coastal and interior sage scrub, chaparral, and oak woodland habitats because of their widespread occurrence. Coniferous forest communities of pines and firs at higher elevations (above about 5,000 feet) are not within the scope of this book.

We give both common names and scientific, or Latin, names for all plants. Common names are more familiar but are often very local, with different names used in different areas. Three or four books may each use a different common name for the same species. Latin names are more stable, but even these are subject to revision as scientists study individual species and develop new understandings of relationships. The Latin names used here follow those of *The Jepson Manual* (Hickman 1993), the most up-to-date summary of the California flora. Flowering periods listed indicate the earliest and latest month in which flowering would be expected to occur. Actual flowering dates in any given year, of course, are subject to seasonal patterns of rainfall.

Beaches and Dunes

Beaches and associated dunes are formed along the coast by the combined actions of ocean currents and erosion on the land. Streams and rivers transport large quantities of sands and gravels to the coast, and alongshore currents deposit these sediments to form beaches. Long sandbars often form at the mouths of bays. Such bars can be seen off the Newport Peninsula in Orange County and the Coronado Peninsula in San Diego. More typical in Southern California, however, are long ribbons of sandy beach, such as those of the Malibu and San Diego coasts. These broad beaches, interrupted only occasionally by durable rocky headlands, are formed by sediments produced locally from the erosion of coastal bluffs. Where the coastal bluffs are composed of conglomerate rock or hard sandstones, however, small cove beaches are produced, such as those of Laguna Beach.

Once beach sands are deposited, winds and waves push these sediments landward, building up dunes that often move with the seasons (pls. 5, 6). Beach sands and dunes are influ-

Plate 5. Sand verbena, beach morning-glory, silver beach bur, and other dune species on the Guadelupe Dunes, Santa Barbara County.

Plate 6. Coastal dune community at the Ballona Wetlands, Los Angeles County, with dune bush lupine, sand verbena, beach evening primrose, and other dune species.

enced strongly by winter storms from the northwest, which move sand southward along exposed beaches. Summer storms more typically originate from the south, pushing sand back northward along individual beaches.

Beach and dune habitats pose serious problems for plant survival. Sandy soils are highly porous and thus hold little water near their surface, resulting in water stress for shallow-rooted plants. Deeper roots provide some advantage, but only to a point, because saltwater typically extends underneath the freshwater that underlies dunes and beaches. Water stress increases as bright sun dries soils and heats leaves, causing higher rates of water loss. Winds deposit ocean salts and sand-blast plant tissues with fine-grained particles. Potentially the most serious problem is that unstable beach sands and dunes can move rapidly, either burying plants or eroding the sand covering their roots.

The most colorful of the common dune plants are the sand verbenas (pls. 7–9): red sand verbena *(Abronia maritima)*, yellow sand verbena *(A. latifolia)*, and pink sand verbena

Plate 7. Red sand verbena (*Abronia maritima*, Nyctaginaceae), February to October.

Plate 8. Yellow sand verbena (*Abronia latifolia*, Nyctaginaceae), May to October.

(A. umbellata). These prostrate and trailing herbs with opposite leaves and succulent stems have a notable characteristic: they are covered with tiny glandular hairs that exude a sticky substance to which sand grains adhere tightly. This sandy covering may protect the plants from sandblasting. Thick fleshy roots penetrate deeply into the sand to tap underground pools of water; they also enable the plants to survive burial by providing stores of carbohydrates for the growth of new stems. Another common dune species with succulent trailing stems and a deep fleshy root system is beach morning-glory *(Calystegia soldanella)*. Its heart-shaped leaves are a dark, vibrant green (pl. 10).

Many beach and dune plants have distinctly silvery-colored foliage. This trait may help to reflect some of the in-

Plate 9. Pink sand verbena (*Abronia umbellata*, Nyctaginaceae), all year.

Plate 10. Beach morning-glory (*Calystegia soldanella*, Convolvulaceae), April to August.

tense rays of the sun, just as sunblock protects us from sunburn at the beach. One characteristic dune plant with this trait is silver beach bur *(Ambrosia chamissonis)*, a low semi-woody herb that forms loose mats over the sand (pl. 11). This plant has the unusual characteristic of highly variable leaf forms ranging from simple to highly lobed. No one knows why this variability exists. Barefoot sunbathers in the vicinity of this species quickly become aware of its method of seed dispersal, a very spiny fruit that sticks tenaciously to animal feet. Another species showing striking silver foliage is beach evening primrose *(Camissonia cheiranthifolia)*, an herb with a basal rosette of fleshy leaves up to eight inches across (pl. 12). The rosette typically produces a series of sprawling prostrate stems tipped by large yellow flowers up to two inches or more

Plate 11. Silver beach bur (*Ambrosia chamissonis*, Asteraceae), July to November.

in diameter. Less common in our area, but striking in appearance, is dune bush lupine *(Lupinus chamissonis)*, a low woody species with foliage covered with long silky hairs (pl. 13).

Annual plants are generally not common on dunes because of the severe growing conditions. One common exception, however, is coast weed *(Amblyopappus pusillus)*, an erect species with sticky, aromatic foliage (pl. 14).

Several of the most common species encountered along our coastal beaches and dunes are invaders native to other

Plate 12. Beach evening primrose (*Camissonia cheiranthifolia*, Onagraceae), April to August.

Plate 13. Dune bush lupine (*Lupinus chamissonis,* Fabaceae), March to July.

Plate 14. Coast weed (*Amblyopappus pusillus,* Asteraceae), March to June.

parts of the world. Sea rocket *(Cakile maritima)*, a succulent-leaved annual that sprouts along open sandy beaches, gets its name from its distinctive seed capsule that resembles a two-stage rocket (pl. 15). This capsule has the ability to float for extended periods on the ocean surface, without damage, until storm waves deposit it on an upper beach area, where it readily germinates. Iceplant *(Carpobrotus edulis)*, also called hottentot-fig, was introduced into California in the nineteenth century from South Africa and was widely used to stabilize coastal dunes (pl. 16). Unfortunately, the dense trailing mats of this succulent plant choke out native species. Its value as a dune stabilizer is limited because its relatively shallow roots are readily eroded by winter storms. Less aggressive is another introduced species of iceplant called sea-fig *(C. chilensis)* (pl. 17). This Chilean species has rose magenta flowers and succulent leaves that are distinctly rounded-triangular in cross section, in contrast to the yellow flowers and sharply triangular leaf cross section of the hottentot-fig.

Dune habitats were once common along much of the

Plate 15. Sea rocket (*Cakile maritima*, Brassicaceae), not native, June to November.

Plate 16. Iceplant (*Carpobrotus edulis*, Aizoaceae), not native, April to October.

Plate 17. Sea-fig (*Carpobrotus chilensis*, Aizoaceae), not native, April to September.

Southern California coast, but continuous development pressure and urbanization over the past century dramatically reduced their extent. Local areas of relatively intact beach and dune communities are found along some of the less developed coasts of Santa Barbara and Ventura Counties, but only very small, patchy remnants of good beach dunes remain in Los Angeles County. In the Pleistocene, a broad coastal belt of young dunes and older stabilized dunes extended from the coast up to several miles inland from south of Santa Monica to the Palos Verdes Peninsula. Virtually all of this area has been urbanized. The least disturbed examples of coastal dune habitats in San Diego County occur on the Coronado Peninsula southward to the Mexican border.

Coastal Bluffs and Terraces

Marine cliffs, cut by waves into terraces and later uplifted by geological pressures over thousands of years, now form coastal bluffs and terraces, which provide another distinctive habitat along the Southern California coast (pl. 18). As many as 20 distinct beach terraces, resulting from past changes in sea level, can be recognized in some areas, most notably on the Palos Verdes Peninsula and the Punta Banda Peninsula in northwestern Baja California. Older beach terraces typically support coastal sage scrub communities, as described in the next chapter. Younger terraces, however, often are home to distinctive assemblages of coastal species adapted to tolerate mildly saline soils and frequent high winds.

Plate 18. Marine terraces at Point Dume, Los Angeles County.

Plate 19. Mock heather (*Ericameria ericoides*, Asteraceae), August to November.

Several shrub species characterize coastal bluff and beach terrace habitats. Mock heather *(Ericameria ericoides)* is a low shrub with tiny linear leaves that enters the northern portion of our region (pl. 19). It is most apparent on the dunes in fall, when its yellow flowers are produced abundantly. Coast goldenbush *(Isocoma menziesii)* maintains a prostrate growth form, unlike the upright shrubby form *(I. menziesii* var. *menziesii)* of this species that is common in coastal sage scrub habitats (see pl. 49). Sea-cliff buckwheat *(Eriogonum parvifolium)* is a low spreading shrub of coastal bluffs and stabilized dunes (pl. 20). Its leaves are dark, shiny green above and densely wooly and white below. Also common on sandy bluffs is the coastal dudleya *(Dudleya caespitosa)*, with a basal rosette of shiny green succulent leaves (pl. 21). Although easy to miss most of the year, the coastal dudleya sends up glorious stems of reddish to yellow flowers in late spring. This is just one of a diverse group of dudleya species ranging along the coast of Southern California and northwestern Baja California. Other species are discussed in chapters 3 and 12.

Plate 20. Sea-cliff buckwheat (*Eriogonum parvifolium,* Polygonaceae), all year.

Plate 21. Coastal dudleya (*Dudleya caespitosa,* Crassulaceae), April to July.

Several dull-looking gray shrubs called saltbushes are commonly encountered on coastal terraces and stabilized dunes. These shrubs have inconspicuous male and female flowers borne separately in summer and owe their success in coastal habitats to a high tolerance of moderately saline soils. The most commonly encountered native species are quail bush (*Atriplex lentiformis),* a large shrub reaching three to nine feet in height (pl. 22), and Watson's saltbush *(A. watsonii),* a trailing species forming tangled mats (pl. 23). Several other saltbushes grow along the coast, including both native and nonnative species. Perhaps the most widespread introduced species is Australian saltbush *(A. semibaccata),* which, as its common name indicates, is native to Australia (pl. 24). This is a trailing, prostrate shrub most typically encountered in disturbed areas of coastal terraces or salt marshes.

Plate 22. Quail bush (*Atriplex lentiformis,* Chenopodiaceae), August to October.

Plate 23. Watson's saltbush (*Atriplex watsonii,* Chenopodiaceae), March to October.

Several iceplants, members of the large South African iceplant family, have become aggressive invaders of coastal bluffs and beach terraces. One of these is crystalline iceplant *(Mesembryanthemum crystallinum)*, an annual that can germinate

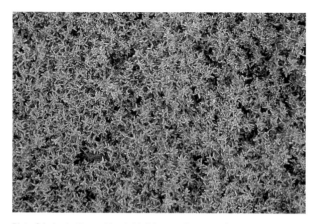

Plate 24. Australian saltbush (*Atriplex semibaccata,* Chenopodiaceae), not native, April to December.

Plate 25. Crystalline iceplant (*Mesembryanthemum crystallinum,* Aizoaceae), not native, March to October.

in massive numbers across coastal terraces, particularly disturbed areas with low levels of salinity (pl. 25). The species name of this small iceplant comes from the appearance of the large bladderlike cells on the leaf surface, which are able to se-

Plate 26. Finger mesemb (*Malephora crocea,* Aizoaceae), not native,
April to July.

quester salts that would otherwise be toxic to the plant. This
species has invaded extensive areas of disturbed coastal terraces of northwestern Baja California. Another shrubby iceplant established in more localized areas is finger mesemb
(Malephora crocea), which forms dense mats with gray green
succulent leaves along coastal bluffs and roadsides (pl. 26).
This species is particularly aggressive in areas with fog and
strongly maritime conditions, such as the Channel Islands
and the Palos Verdes Peninsula.

Characterizing Sage Scrub

Coastal sage scrub and the closely related interior sage scrub are plant communities dominated by drought-deciduous shrubs. Drought-deciduous means that these species lose their foliage in summer with the onset of drought conditions, then grow new leaves the following fall when rains replenish soil moisture. This is a very different form of leaf loss than that of winter-deciduous oaks and riparian trees, which lose their leaves in late fall and form new leaves in early spring. There is a third form of sage scrub, termed maritime succulent scrub, that is characterized by a significant presence of succulent plants. This maritime succulent scrub is characteristic of northwestern Baja California but extends into California in a few places, as discussed below.

Why are these sage scrub communities dominated by species that are deciduous in summer, while chaparral shrubs keep virtually all of their leaves throughout the year? The answer lies with the relatively dry conditions favored by coastal sage scrub and interior sage scrub. These communities largely occur in semi-arid habitats along the coast, and again inland in the rain shadow margins of the coastal hills and on the edges of the desert. These habitats receive no more than 10 to 12 inches of rain each year and have a predictably long dry period that typically extends for six months or more. Under such conditions it is very expensive, in terms of carbon resources, for shallow-rooted plants such as these to maintain active leaves during the summer, when water is not available to them. The ecological strategy of these shrubs is to produce leaves with high rates of photosynthesis during those months when moisture is available, and then drop these leaves during the drought season. As long as the photosynthetic rates are high, carbon resources are used more efficiently in growing new leaves than in maintaining old ones.

Most of our desert shrubs follow exactly the same drought-deciduous pattern.

Chaparral shrubs have a very different ecological strategy. Their evergreen leaves are potentially active at any time during the year. This strategy is effective when soil moisture is available because of higher rainfall and shorter summer drought periods, or when deep roots can tap groundwater resources. The tradeoff, however, is that thick, leathery leaves inherently have much lower rates of photosynthesis than do thinner deciduous leaves. Low rates of photosynthesis throughout the year can provide the same amount of stored carbon as high rates of photosynthesis for half of the year.

Sage scrub communities are structurally distinct from chaparral in ways beyond their drought-deciduous behavior. The typical height of sage scrub shrubs is only about three or four feet, compared to four to eight feet or more in chaparral. Walking in sage scrub communities is relatively easy because the shrubs are less densely packed and the canopies lack the stiff woody branches that make mature chaparral virtually impenetrable. Still, in some cases evergreen woody shrubs can form an important component of dominance in coastal sage scrub communities. Evergreen shrub species are able to thrive along the coast and inland near the desert in habitats where their root systems can penetrate deeply into the soil to exploit extra water resources during the dry summer months, even when total rainfall is low.

The separation between coastal sage scrub and chaparral communities along the coast is relatively well defined in San Diego County but often much less clear in areas such as the Santa Monica Mountains (pl. 27). Chaparral communities with evergreen shrubs can establish on coastal slopes if the underlying geology allows their roots to penetrate deeply into the soil. Although these conditions would also be fine for coastal sage scrub species, the larger sizes and dense canopies of chaparral shrubs allow them to dominate over the smaller

Plate 27. Ocean-facing coastal sage scrub community with California encelia, California sagebrush, and black sage.

drought deciduous shrubs. Similarly, coastal sage shrubs can readily survive on rocky slopes and disturbed sites such as landslides within chaparral stands as long as they are not in direct competition with chaparral shrubs.

Because sage scrub shrubs lack woody root crowns or burls, their response to fire is very different from that of chaparral shrubs. Light fires in sage scrub may burn off surface foliage but still allow a limited amount of resprouting from the plants' bases. Resprouting does not occur, however, after an intense fire that burns off most of the aboveground tissues. These shrubs must be replaced by seeds from outside the area as they lack soil seed pools. The differing responses to fire may explain why most sage scrub shrubs have seeds that are wind dispersed, whereas most evergreen chaparral shrubs have fruits dispersed by birds or mammals. Wind dispersed seeds allow the rapid recolonization of burned areas of sage scrub.

Although the name coastal sage scrub is commonly used

broadly to describe a series of related plant communities, it seems appropriate to recognize three main forms. Coastal sage scrub occurs along most of our coastal region, interior sage scrub at the eastern margin of the Los Angeles Basin and in inland areas near the Peninsular Ranges, and maritime succulent scrub in a few areas of southern San Diego County and southward into northwestern Baja California. A related community is maritime chaparral, a coastal community formed of a few specific evergreen shrubs mixed with coastal sage scrub.

Coastal Sage Scrub

The habitat we call coastal sage scrub extends in our region from Ventura County southward along the lower coastal slopes of the Santa Monica Mountains (pl. 28), and along the coastal hills and slopes of Orange and San Diego Counties. This community combines the common sage scrub shrubs

Plate 28. Coastal sage scrub in Tuna Canyon, Santa Monica Mountains, with laurel sumac, black sage, and buckwheat.

with a diverse assemblage of species whose range is largely restricted to sites close to the coast. Increases in species richness from north to south have led some scientists to divide these communities into a northern form termed Venturan sage scrub and a southern form called Diegan sage scrub.

Relatively few species form the primary cover of coastal sage scrub. Generally the most visible of these are widespread species able to quickly colonize disturbed sites. Their dominance today reflects the long history of human disturbance in coastal habitats, such as changes in fire frequency and intensity, cattle grazing, and development. Many of these species are not restricted to sage scrub communities but occur also in open or disturbed areas of chaparral. A good example is deerweed *(Lotus scoparius)*, discussed in more detail in chapter 5 (see pl. 151). This plant is a common drought-deciduous shrub able to quickly colonize open areas of chaparral.

Two common coastal sage scrub species are in the sunflower family (Asteraceae). California sagebrush *(Artemisia californica)*, the more abundant of these, is easy to recognize, with its finely dissected gray green foliage and distinctive

Plate 29. California sagebrush (*Artemisia californica*, Asteraceae), August to December.

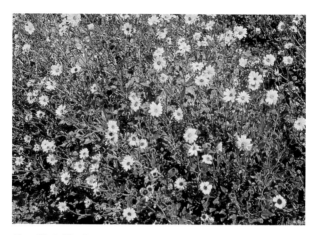

Plate 30. California encelia (*Encelia californica,* Asteraceae), February to June.

pungent odor (pl. 29). The early Spanish settlers equated its smell with medicinal values and used sagebrush teas and salves as cures for almost every ailment. The second shrub is California encelia *(Encelia californica)*, a semiwoody shrub characterized by large, erect flower heads with yellow rays and a dark brown disk (pl. 30). Dense populations of this species color many steep coastal slopes bright yellow in early spring. Like California sagebrush, it owes much of its success to its ability to effectively colonize disturbed sites.

Several species of sage are important components of coastal sage scrub communities. All are members of the mint family (Lamiaceae), which characteristically has two-lipped flowers, a square stem in cross section, and fragrant foliage. The most widespread is black sage *(Salvia mellifera)*, a shrubby species with narrow, dark gray green leaves (pl. 31). Besides being important in coastal sage scrub throughout Southern California, black sage is also a common associate of chamise *(Adenostoma fasciculatum)* in the transition from chamise chaparral to interior sage scrub on the inner side of

Plate 31. Black sage (*Salvia mellifera*, Lamiaceae), April to July.

the Santa Monica Mountains and the eastern margin of the chaparral in Riverside and San Diego Counties. A similar species in general form is purple sage *(S. leucophylla)*, a shrub with broader, whitish gray leaves densely covered by short woolly hairs (pl. 32). Another characteristic that separates purple sage from black sage is the presence of long stamens (the male reproductive organs) that extend far beyond the petals. Although both species are widespread and common, the two rarely occur together in a stand. White sage *(S. apiana)*, a large but less woody shrub (pl. 33), is also widespread but occurs in lower densities than black sage or purple sage. White sage is clearly distinguished by the woolly white hairs that cover both stems and leaves.

One widespread species that can dominate stands of coastal sage scrub on disturbed sites is California buckwheat

Plate 32. Purple sage (*Salvia leucophylla,* Lamiaceae), May to July.

Plate 33. White sage (*Salvia apiana,* Lami-aceae), April to July.

Plate 34. California buckwheat (*Eriogonum fasciculatum,* Polygonaceae), May to October.

(Eriogonum fasciculatum). This is a low shrub two to three feet in height, with dense heads of tiny pinkish flowers (pl. 34). The tiny linear leaves occur in bundles along the stems. The upper leaf surface is dark green, with a downward-rolled margin that partially hides a white woolly lower surface. Unlike the majority of coastal sage shrubs, California buckwheat is semievergreen, keeping the majority of its leaves throughout the summer. This species and the sages described above are important plants for bee culture and produce a valued honey.

One species of evergreen shrub must be considered a typical and important component of coastal sage scrub communities. Lemonadeberry *(Rhus integrifolia)* occurs at lower elevations, mostly below 1,000 feet, on the ocean-facing side of the coastal ranges (pl. 35). This species may well have once been much more common along the coast, for many stands have been cleared away for urban and recreational developments. Large expanses of lemonadeberry, however, can still be found locally in many coastal areas, such as parts of the Palos

Verdes Peninsula and the less developed canyons of Malibu and the San Diego County coast. The common name comes from the sticky pulp on the outer surface of mature fruits, which has a lemony flavor and can be used to make a refreshing beverage. Another woody evergreen shrub common in coastal sage scrub is laurel sumac *(Malosma laurina)*, a species more typical of chaparral and described in chapter 4 (see pl. 104). Both of these species, and particularly the latter, are sensitive to cold temperatures at higher elevations and inland sites.

Beyond these common shrubs, a number of less common ones are widespread in coastal sage scrub. Golden yarrow *(Eriophyllum confertifolium)* is a low semiwoody member of the sunflower family that is abundant in open areas away from the coast (pl. 36). The beautiful golden yellow flower clusters at the ends of branches make this a striking plant in spring. The young plants have a dense matting of woolly white hairs along their surface. In summer, golden yarrow dies back to its woody base and disappears from view until the first rains of fall. Also spectacular when flowering is the wishbone bush *(Mirabilis californica)*, a low semishrub with showy red purple flowers (pl. 37). The strange common name of this species is difficult to understand in spring when the

Plate 35. Lemonade-berry (*Rhus integrifolia*, Anacar-diaceae), February to May.

Plate 36. Golden yarrow (*Eriophyllum confertifolium*, Asteraceae), April to August.

Plate 37. Wishbone bush (*Mirabilis californica*, Nyctaginaceae), December to June.

plants are flowering. It becomes more obvious in late summer, however, when much of the aboveground foliage dies, leaving forked branches resembling turkey wishbones. California croton *(Croton californicus)* is an undistinguished semiwoody shrub up to two feet in height that is common on sandy flats near the coast (pl. 38).

Bladderpod *(Cleome arborea)* is an erect and branched shrub up to six feet or more in height. It is drought tolerant

Plate 38. California croton (*Croton californicus*, Euphorbiaceae), March to October.

Plate 39. Bladderpod (*Cleome arborea,* Capparaceae), all year.

and possesses thin, evergreen leaves having three leaflets half an inch to an inch long, with a strong, pungent—some say ill-smelling—odor (pl. 39). Bladderpod has an unusual pattern of distribution; it is common on bluffs and hills near the coast

Plate 40. Thick-leaf yerba santa (*Eriodictyon crassifolium*, Hydrophyllaceae), April to June.

and also in the western edges of the Sonoran and Mojave Deserts. The common name comes from its inflated bladder-like fruits. Bladderpod is one of the few species in our flora that can be found flowering in any month of the year.

Thick-leaf yerba santa *(Eriodictyon crassifolium)* can also be found on shallow rocky soils (pl. 40) in both open chamise coastal sage scrub and chaparral communities discussed in chapter 4. Although truly woody only at its base, our yerba santa can reach four to six feet in height. Its two- to four-inch leaves with scalloped or toothed margins are evergreen and have densely matted woolly hairs growing over both leaf surfaces. A leaf added to a cup of boiling water makes a soothing tea. The name yerba santa was given to this plant by the early mission padres, who were greatly impressed when Native Americans demonstrated its curative powers for respiratory infections, fevers, and sores. Trask's yerba santa *(E. traskiae)* is a low-growing shrub similar in appearance (pl. 41), with an unusual distribution; it occurs in the Santa Inez Mountains of Santa Barbara County and on Santa Catalina Island. It is dif-

Plate 41. Trask's yerba santa (*Eriodictyon traskiae*, Hydrophyllaceae), May to June.

Plate 42. Prickly phlox (*Leptodactylon californicum*, Polemoniaceae), March to June.

ferentiated from thick-leaf yerba santa by the presence of glandular hairs on the flower petals and sepals.

Prickly phlox *(Leptodactylon californicum)* is a small, openly branched shrub two to three feet in height with woolly stems covered in bunches of prickly needlelike leaves (pl. 42).

The small stature and scattered occurrence of these plants make them largely invisible until spring, when they burst out with masses of small pink flowers that can be seen at considerable distances. Another semiwoody plant that seems invisible for much of the year until it bursts into bloom in summer is the California brickellbush *(Brickellia californica),* which reaches three to four feet in height but contains little woody tissue. The ovate leaves have scalloped edges, a heart-shaped base, and woolly hairs on both surfaces (pl. 43). The cream-colored flower heads give off a pleasant odor at night.

Plate 43. California brickellbush (*Brickellia californica,* Asteraceae), August to October.

Bush monkeyflower *(Mimulus aurantiacus),* often called sticky monkeyflower, is a low shrub common in disturbed areas of coastal sage scrub as well in other communities, including chaparral and live oak woodland. The name monkey-flower comes from the two-lipped flowers, which look, from the front, like small faces (pl. 44). Bush monkeyflower is typically two to four feet in height and easily characterized by its opposite leaves that are sticky to the touch. The flowers, borne in pairs along the stems, vary in shape. Variation in color from yellow and orange yellow to salmon may reflect an evolving change from bee to hummingbird pollination. An interesting sensitivity to touch can be seen in the stigma (the female reproductive organ), which has two prominent lips spread horizontally. If you lightly touch the stigma with a small twig, the

Plate 44. Bush monkeyflower (*Mimulus aurantiacus*, Scrophulariaceae),
March to July.

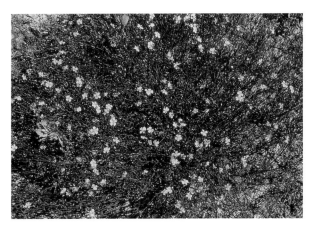

Plate 45. Rock-rose (*Helianthemum scoparium*, Cistaceae), March to
June.

two lips will close after a second or two. If no pollen is deposited, the lips will reopen within an hour.

Rock-rose *(Helianthemum scoparium)* is a low, multi-
stemmed semiwoody shrub (pl. 45) that occurs most com-

monly at lower elevations near the coast. Like many coastal sage scrub species, it can also be found in open stands of chamise chaparral. The tiny linear leaves are lost in summer, making the plant look almost like a bunchgrass at a distance. Clusters of small yellow flowers appear in spring and summer. Rock-rose is a close relative of the genus *Cistus* from the Mediterranean Basin, whose species are widely planted in Southern California gardens.

Ashyleaf buckwheat *(Eriogonum cinereum)* is a straggly semievergreen shrub two to four feet in height with small gray green leaves (pl. 46). It is abundant near the coast on rocky slopes and ridges with thin soil in the northern part of our region. It is particularly abundant in the Santa Monica Mountains, where it colonizes steep canyon walls.

Three more shrubby members of the sunflower family found in coastal sage scrub are weedy opportunists in their ecology, although they are native species. They also share traits of drought resistance that allow them to maintain their leaves throughout the summer months, despite an absence of the deep roots and leathery leaf texture typical of evergreen chaparral shrubs. Coyote brush *(Baccharis pilularis)* is a large shrub with small, toothed leaves (pl. 47) that is common in open areas near the coast and is found occasionally in grasslands and chaparral. Sawtooth golden-

Plate 46. Ashyleaf buckwheat (*Eriogonum cinereum,* Polygonaceae), June to December.

Plate 47. Coyote brush (*Baccharis pilularis*, Asteraceae), August to October.

Plate 48. Sawtooth goldenbush (*Hazardia squarrosa,* Asteraceae), August to October.

bush *(Hazardia squarrosa)* is a small shrub up to three feet in height. It has brittle stems lined with clasping leaves that have sharp teeth along their margins (pl. 48). Like coyote brush, this is a colonizing species that readily establishes in disturbed sites in coastal sage scrub, chaparral, and grasslands. Similar in general appearance but with foliage sticky to the touch and lacking sharp-toothed leaf margins is coast goldenbush *(Isocoma menziesii,* formerly known as *Haplopappus venetus),* which favors sandy soils both near the coast and in-

Plate 49. Coast golden-bush (*Isocoma menziesii* var. *menziesii*, Asteraceae), April to December.

land (pl. 49). All three of these drought-tolerant species flower from late summer into fall, the driest period of the year.

Several cactus species are common associates of coastal sage scrub and become increasingly abundant as you move southward along our coast. Coast prickly-pear *(Opuntia littoralis)* is the most common of these cacti (pl. 50). It frequently forms large shrubby thickets that dominate ground cover over extensive areas of rocky slopes. The joints of this species are longer than they are wide and have nodes with long white spines. Although fires may kill small prickly-pears, only the outer margins of large thickets are harmed, and thus the plants survive. A related species is tall prickly-pear *(O. oricola)*, a stalked species that can be either shrubby or treelike but never forms extensive thickets like those of coast prickly-pear (pl. 51). Tall prickly-pear occurs in scattered locations along the Southern California coast but is perhaps more typical of the maritime succulent scrub described below. It can be distinguished from coast prickly-pear by the almost perfectly

Plate 50. Coast prickly-pear (*Opuntia littoralis*, Cactaceae), June to July.

round joints and by spines that are yellow and, on close examination, distinctly flattened.

Coast cholla *(O. prolifera)* differs from the prickly-pears in having fleshy cylindrical joints along branched stems that reach to four to six feet in height (pl. 52). This species is rare along the coast of the Santa Monica Mountains but becomes

Plate 51. Tall prickly-pear (*Opuntia oricola*, Cactaceae), June to July.

Plate 52. Coast cholla (*Opuntia prolifera*, Cactaceae), April to June.

more common around the margins of the Palos Verdes Peninsula. It is abundant in coastal sage scrub and maritime succulent scrub in southern San Diego County and northwestern Baja California. All species of *Opuntia*, including these three, have two kinds of spines. The larger spines are relatively easy to see and thus avoid, but the areoles around the base of the clusters of spines on the stems and fruits contain large numbers of tiny spines called glochids, which are heavily barbed and are thus important to avoid. These almost invisible spines quickly embed themselves in fingers and hands and are very difficult to remove.

A final common associate of coastal sage scrub, as well as of open stands of chamise chaparral, is chaparral yucca *(Yucca whipplei),* also known as our lord's candle. This relative of the century plants, or agaves, forms a rosette of two- to three-foot-long bayonetlike leaves with sharp tips (pl. 53). In late spring each mature rosette rapidly elongates into a tall flowering stalk

Plate 53. Chaparral yucca (*Yucca whipplei,* Agavaceae), April to June.

eight to 10 feet in height topped by a dense cluster of cream-colored flowers. During its active period of elongation, this flowering stalk may grow a remarkable four to six inches each day. Soon after its single flowering period, the rosette dies. This seems to be the ultimate in maternal altruism!

Interior Sage Scrub

The eastern half of the Los Angeles Basin, western Riverside County, and adjacent northeastern San Diego County support a community that is very similar to coastal sage scrub in general structure (pl. 54) but is, overall, lower in species richness. To call this coastal sage scrub would seem rather silly as it occurs far from the coast, and thus we use the name interior sage scrub. This community has also been called Riversidian sage scrub.

Plate 54. Interior sage scrub in western Riverside County.

The majority of common coastal sage scrub species can be found in interior sage scrub, including deerweed, California sagebrush, California encelia, black sage, white sage, California buckwheat, golden yarrow, and chaparral yucca. Several species, however, do not make the jump to the interior, most notably lemonadeberry and purple sage among the common shrubs, and a number of other shrubs limited to the immediate coast.

Because it is close to the inland deserts, interior sage scrub contains several important shrubs and succulents not found commonly in coastal sage scrub. One of these is brittlebush (*Encelia farinosa*), a species more typical of the warmer and drier Sonoran Desert. Brittlebush is closely related to California encelia but differs sharply in appearance because of the dense woolly hairs that cover the leaves and stems (pl. 55). Leaves of brittlebush are larger and less hairy in winter because they form under conditions of good water availability, but they gradually become smaller, thicker, and more hairy as spring moves into summer. These seasonal changes in leaf morphology help adapt brittlebush to its dry summer habi-

Plate 55. Brittlebush (*Encelia farinosa*, Asteraceae), March to May.

Plate 56. Cane cholla (*Opuntia parryi*, Cactaceae), May to June.

tat. Another semidesert species is cane cholla *(Opuntia par-ryi),* an upright or rarely prostrate shrubby cactus with cylin-drical joints (pl. 56). Cane cholla occasionally reaches to the coast, as at Cabrillo National Monument on the Point Loma Peninsula near San Diego.

Maritime Succulent Scrub

The third form of sage scrub, maritime succulent scrub, occupies the more arid southern coastal areas of our region. Although a number of characteristic coastal sage scrub species are present, maritime succulent scrub includes prominent succulent species as well as several notable evergreen shrubs. Maritime succulent scrub is the dominant form of sage scrub along the coastal areas of Baja California from Ensenada northward to Tijuana and enters into a few areas of Southern California. It can best be seen at Cabrillo National Monument and Torrey Pines State Park (pl. 57) but occurs as well on dry south-facing coastal slopes on San Clemente and Santa Catalina Islands. Many of the hardier and more fire-resistant succulent species from this community extend northward, with scattered occurrences in coastal sage scrub and interior sage scrub habitats.

Maritime succulent scrub almost certainly has the greatest species richness of any of the sage scrub communities, despite the arid nature of its habitat. One factor responsible for this added diversity, much of which is in succulent and other sensitive plant species, is that these coastal stands are largely free of fire.

Shrubs such as California encelia, California sagebrush, lemonadeberry, deerweed, and California buckwheat, all widespread in coastal sage scrub, are also common in maritime succulent scrub. Black sage, however, is increasingly replaced in southern San Diego County by the closely related Munz sage *(Salvia munzii),* a low shrub with small gray green leaves and similar ecological requirements (pl. 58). In addition, maritime succulent scrub includes two evergreen species also found in maritime chamise chaparral, which is described in chapter 4. The first is bushrue *(Cneoridium dumosum),* a low evergreen shrub that occurs as far north as Laguna Beach. It is a member of the citrus family, a relationship

Plate 57. Maritime succulent scrub at Torrey Pines State Park, San Diego County, with California boxthorn, coast prickly-pear, lemonade-berry, and coast barrel cactus.

Plate 58. Munz sage (*Salvia munzii*, Lamiaceae), February to April.

Plate 59. Bushrue (*Cneoridium dumosum,* Rutaceae), November to March.

Plate 60. Jojoba (*Simmondsia chinensis,* Simmondsiaceae), March to May.

made evident by the smell of the resin-dotted berries and foliage (pl. 59). The second is jojoba *(Simmondsia chinensis),* a species more typical of inland deserts (pl. 60). Jojoba has separate male and female plants, with the females widely cultivated (with scattered male plants for fertilization) for the valuable oil contained in the fruits.

Several shrub species of maritime succulent scrub favor rocky flats and slopes close to the ocean. California boxthorn *(Lycium californicum)* is a dense and intricately branched woody shrub that may reach three to six feet in height. This shrub is easily recognized by its spiny branches and small, cylindrical succulent leaves (pl. 61). As the common name implies, this species occurs both along the coast and inland in

Plate 61. California box-thorn (*Lycium californicum*, Solanaceae), March to July.

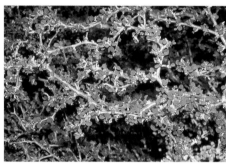

Plate 62. Cliff spurge (*Euphorbia misera,* Euphorbiaceae), January to August.

the desert. Cliff spurge *(Euphorbia misera)* is a low summer-deciduous shrub with succulent gray stems that contain a milky sap (pl. 62). This sap is highly toxic and should be kept from contacting the skin or eyes.

The succulents that give maritime succulent scrub its name include a diverse group of cacti, as well as agaves and other small succulents. The three coastal species of prickly-pear and cholla described above are all significant members of this community. Two other large succulents that are more typical of northwestern Baja California are also important, but only in the southernmost parts of our region. The first of these is a beautiful cactus called golden cereus *(Bergerocactus emoryi)*, which forms shrubby clumps of upright cylindrical stems two

Plate 63. Golden cereus (*Bergerocactus emoryi*, Cactaceae), May to June.

to three feet in height (pl. 63). This species occurs in scattered coastal locations from Del Mar southward and on San Clemente and Santa Catalina Islands. A surprising trait of golden cereus is that it can resprout from its base after fire, one of very few cactus species to do so. The other large succulent is coastal agave *(Agave shawii)*, a rare species in our region but one abundant below the border in northwestern Baja California. Coastal agave forms clusters of huge rosettes with sharp spines along the succulent leaves (pl. 64). The tall flowering stalks grow rapidly to heights of six to eight feet or more.

In addition to these large succulents, maritime succulent scrub commonly includes a number of smaller succulent species. Two of these are cacti that, like their larger brethren described above, are more typical of coastal areas of northwestern Baja California. One is the coast barrel cactus *(Ferocactus viridescens)*, a small and ribbed cylindrical species six to 10 inches in height (pl. 65). The other is the fishhook cactus

Plate 64. Coastal agave (*Agave shawii,* Liliaceae), September to May.

Plate 65. Coast barrel cactus (*Ferocactus viridescens,* Cactaceae), May to June.

Plate 66. Fishhook cactus (*Mammillaria dioica,* Cactaceae), February to April.

(Mammillaria dioica), an even smaller cylindrical species two to four inches in height, with spines barbed at the end to resemble fishhooks (pl. 66).

The other common group of small succulents consists of dudleya species *(Dudleya),* also called live-forever, which are commonly present in open areas of maritime succulent scrub. *Dudleya* comprises about 45 diverse species, with its species richness centered along the coast of Southern California and northwestern Baja California, where these plants are common on coastal flats, rock outcrops, and open areas of chamise chaparral. The most commonly encountered species are lance-leaved dudleya *(D. lanceolata),* with slender green leaves (pl. 67), and chalk dudleya *(D. pulverulenta),* with chalky white leaves (pl. 68), but many other less common species can be found locally. Under the conditions prevalent in the maritime succulent scrub communities of northwestern Baja California, species such as Britton's dudleya *(D. brittonii)* (pl. 69) provide a significant component of plant cover.

Plate 67. Lance-leaved dudleya (*Dudleya lanceolata,* Crassulaceae), May to July.

Plate 68. Chalk dudleya (*Dudleya pulverulenta,* Crassulaceae), May to July.

Plate 69. Britton's dudleya (*Dudleya brittonii,* Crassulaceae), May to July (photograph by Philip Rundel).

Characterizing Chaparral

The shrublands that we call chaparral form the most prominent and widespread plant community in Southern California. Chaparral dominates broad areas of the foothills of the South Coast, Transverse, and Peninsular Ranges of Southern California, particularly on drier or shallow soils where oak woodlands cannot survive (pls. 70, 71). On these slopes chaparral forms dense and virtually impenetrable thickets of woody shrubs anywhere from four to eight feet or more in height. At low elevations near the coast with less rainfall and along the transition to interior deserts, there is commonly a transition from evergreen chaparral communities to the summer-deciduous sage scrub communities described in the previous chapter. With cooler temperatures at higher elevations above about 4,500 to 5,500 feet in the Transverse and Peninsular Ranges, chaparral communities drop out and are replaced by conifer forests.

The name chaparral comes from the Spanish word *chaparro,* which originally was used to describe a cover of shrubby

Plate 70. Mixed chaparral at Gold Creek in the San Gabriel Mountains.

Plate 71. Mixed chaparral community with ceanothus, toyon, and chamise in the Santa Monica Mountains.

evergreen oaks in Spain. Today the word is used broadly to describe our California evergreen shrublands whether or not shrubby oaks are present. A century ago East Coast writers unfamiliar with California sometimes referred to chaparral as the California elfin forest.

Although numerous shrub species are present in most chaparral stands, many of them have evolved such similar growth forms and leaf shapes that it takes some training of the eye to distinguish them. Most chaparral shrubs have a rigid woody branching structure and leathery evergreen leaves that are present throughout the year. They typically have both deep roots that tap pools of water well below the ground and shallow roots that forage for water and nutrients in the upper layers of soil. The deep root system, together with a number of physiological traits that help them tolerate drought stress, allows these plants to remain active throughout the summer months when surface soil moisture is no longer available. Access to soil moisture is critical for evergreen leaves to maintain physiological activity in these dry months.

Plant communities very similar in structure to our chaparral have evolved in each of the five regions of the world where a mediterranean climate exists. These evergreen shrublands are called *matorral* in central Chile, *garrigue* in the Mediterranean Basin, *fynbos* in South Africa, and *kwongan* or *heathland* in Southwestern Australia. The structural similarity of these shrublands does not indicate a common origin or a past connection of these areas. Instead, it results from convergent evolution: environmental stresses presented by mediterranean climate conditions have led to the development of similar morphological and physiological adaptations.

Chaparral communities vary greatly in the composition of their dominant species. This is not surprising, because more than 100 shrub species grow somewhere in chaparral stands in Southern California. Fewer than 40 of these are widespread and common, however, and only a small number are normally present in any local chaparral stand. On north-facing slopes it is common to find stands of chaparral where six to 10 or more shrub species share dominance. At the other extreme, extensive chaparral stands on dry south-facing slopes are virtual monocultures of chamise *(Adenostoma fasciculatum)*. Beyond the simple designations of mixed chaparral and chamise chaparral, it is difficult to characterize consistent forms of chaparral communities.

Chamise Chaparral

Chamise, a shrubby member of the rose family (Rosaceae), is the most abundant and characteristic shrub of the California chaparral (pl. 72). It can be found over the entire range of chaparral communities from southern Oregon down the Coast Ranges and the foothills of the Cascades and Sierra Nevada into Southern California. In our region it occurs on the foothills of the Coast, Transverse, and Peninsular Ranges

Plate 72. Chamise chaparral in western Riverside County.

at elevations up to about 5,000 feet, and it extends southward into Baja California.

With its strong ability to survive drought conditions, chamise thrives on rocky sites with shallow soils and dry south-facing slopes that do not allow the growth and survival of other chaparral shrubs. It has a remarkable ability to reestablish itself after fires, as described in the next chapter. Chamise may also grow on deeper soils or north-facing slopes where water is less limiting, but on these sites it loses its competitive advantage and must compete with a diversity of other species. Of all our chaparral shrubs, chamise is one of the most distinctive and easiest to identify because of its tiny needlelike leaves and peeling bark (pl. 73). Typical stands of chamise reach heights of four to eight feet, although a dwarf form of chamise described below occurs in maritime chamise chaparral in portions of coastal San Diego County and northwestern Baja California.

A striking feature of many chamise stands is the overwhelming dominance of this species on many south-facing slopes, in sharp contrast to the shared dominance of multiple shrub species in most chaparral stands. In late spring the clusters of small white flowers at the branch tips color entire hillsides, giving them a distinctive appearance. As chamise flowers mature and fruits form and are dispersed, the branch tips take on a bronze color from the remaining flower stems and give this color to the landscape.

A second species of *Adenostoma* is also present in Southern California, but much less com-

Plate 73. Chamise (*Adenostoma fasciculatum*, Rosaceae), May to June.

monly than chamise. This is redshank *(A. sparsifolium)*. Redshank is typically found in the more open and arid communities of chaparral on the leeward slopes of the Coast, Transverse, and Peninsular Ranges, where a rain shadow limits water availability and chaparral begins to grade into semi-arid or desert communities. Although it shares the traits of needle-like leaves and peeling bark, redshank looks quite different than chamise. The shrubs are much larger, commonly eight to 15 feet or more in height, and the sticky leaves are longer and thinner than those of chamise and bright green rather than olive green in color (pl. 74). The masses of reddish peeling bark of redshank, the source of its name, also differ from the less dramatic gray peeling bark of chamise.

Although chamise stands are typically low in species diversity, a number of small semiwoody shrubs commonly can be found within or around their margins. Many of these species or their close relatives also occur in the coastal sage scrub of more arid habitats. Two of the most common associ-

Plate 74. Redshank (*Adenostoma sparsifolium,* Rosaceae), July to August.

ates of chamise are species of salvia, with drought-deciduous leaves that are shed in summer. Black sage *(Salvia mellifera)*, discussed in more detail in chapter 3, is a common associate of chamise in the transition from chamise chaparral to interior sage scrub on the inner side of the Santa Monica Mountains and the eastern margin of the chaparral in Riverside and San Diego Counties. Fragrant sage *(S. clevelandii)* is restricted in California to stands of chamise chaparral and coastal sage scrub in San Diego County (pl. 75). Woolly blue-curls *(Trichostema lanatum)*, another strongly pungent member of the mint family (Lamiaceae), is a common shrub in dry stands of

Plate 75. Fragrant sage (*Salvia clevelandii,* Lamiaceae), April to July.

Plate 76. Woolly blue-curls (*Trichostema lanatum*, Lamiaceae), May to August.

chamise chaparral (pl. 76). Redberry *(Rhamnus crocea)* is a widespread evergreen shrub with spiny branchlets that is frequently found in chamise chaparral (pl. 77).

Other species common in open stands of chamise chaparral are also common in the coastal and interior sage scrub that

Plate 77. Redberry (*Rhamnus crocea*, Rhamnaceae), March to April.

were treated in chapter 3. These include California sagebrush *(Artemisia californica)*, California buckwheat *(Eriogonum fasciculatum)*, golden yarrow *(Eriophyllum confertifolium)*, bush monkeyflower *(Mimulus aurantiacus)*, wishbone bush *(Mirabilis californica)*, prickly phlox *(Leptodactylon californicum)*, thick-leaf yerba santa *(Eriodictyon crassifolium)*, and chaparral yucca *(Yucca whipplei)*.

Although herbaceous species are rare in stands of chamise chaparral, one notable exception is soap plant *(Chlorogalum pomeridianum)*, a common bulb plant with long wavy-margined leaves that trail across the ground for two feet or more. The white starlike flowers are scattered in small numbers along a tall naked flowering stem (pl. 78). The flowering stalk of soap plant is typically three to five feet tall, but following fires it reaches twice this height. The common name comes from the presence in the large bulbs of saponins, chemicals that form a soapy lather when mixed with water. One effect of this chemical in solution is to prevent the gills of fish from taking up oxygen from the water around them. Native Americans crushed the bulbs and threw them into small streams as a way to force fish to the surface for easy capture.

Plate 78. Soap plant (*Chlorogalum pomeridianum*, Liliaceae), May to June.

Plate 79. Plummer's mariposa lily (*Calochortus plummerae*, Liliaceae), May to July.

A number of other herbaceous plants with bulbs or bulb-like corms may also be found in open areas within chaparral or coastal sage scrub stands. These species typically occur in grasslands and are thus discussed in chapter 7. One of the mariposa lilies, however, is very clearly a chaparral species. This is Plummer's mariposa lily *(Calochortus plummerae),* a slender branched species with flowers that range from pink and rose to white (pl. 79).

Maritime Chamise Chaparral

Maritime chaparral is an unusual community dominated by a dwarf form of chamise and a mixture of other shrub species (pl. 80). Although dominated by evergreen species and thus best considered a form of chaparral, maritime chamise chaparral intergrades with maritime succulent scrub, as described in chapter 3. Both were once more widespread on coastal mesas of San Diego County and in similar coastal habitats of northwestern Baja California.

The chamise that dominates maritime chaparral in our region is a form that grows to only about two to three feet in height. Occurring with it are wart-stemmed ceanothus *(Ceanothus verrucosus)* (pl. 81), mission manzanita *(Xylococcus bi-*

Plate 80. Southern maritime chamise chaparral at Torrey Pines State Park with mission manzanita, scrub oak, wart-stemmed ceanothus, and small-flowered mountain mahogany.

Plate 81. Wart-stemmed ceanothus (*Ceanothus verrucosus*, Rhamnaceae), March to April.

Plate 82. Mission manzanita (*Xylococcus bicolor*, Ericaceae), December to February.

Plate 83. Mojave yucca (*Yucca schidigera*, Liliaceae), April to May.

color) (pl. 82), and typical coastal scrub species such as black sage and California buckwheat. Stands of this community at Torrey Pines State Park and Cabrillo National Monument in San Diego County contain several unexpected desert shrubs. These include jojoba *(Simmondsia chinensis)* (see pl. 60) and

Mojave yucca *(Yucca schidigera)* (pl. 83). A few elements of this community, in particular dwarf chamise with bushrue *(Cneoridium dumosum)* (see pl. 59) and wart-stemmed ceanothus, can still be found in local areas of the coastal hillsides of South Laguna Beach. Urbanization of the Southern California coast and the beach terraces of Baja California between Tijuana and Ensenada has destroyed much of this habitat, however.

Mixed Chaparral

North-facing chaparral slopes, as well as slopes with other exposures that lack the aridity of chamise chaparral sites, typically support a community called mixed chaparral in which many species may be present. Often six to 10 or more species of evergreen woody shrubs are present, with one or a group of these dominating the canopy cover (pls. 84, 85). These shrubs commonly include scrub oaks, manzanitas, ceanothus, and many other species. Local dominance has given rise to community names such as scrub oak chaparral, ceanothus chaparral, and manzanita chaparral.

Plate 84. Mixed chaparral community in the Santa Monica Mountains with Eastwood manzanita, chaparral pea, and black sage.

Plate 85. Mixed chaparral community in the Santa Monica Mountains with redshank, laurel sumac, ceanothus, and bush monkeyflower.

Although we think of these mixed chaparral shrubs as supremely adapted to our mediterranean climate, we know from the fossil record that most of them evolved in subtropical areas of northern Mexico with a seasonal climate of dry winters and summer rains. Although this was the reverse of what we have today, the adaptations to seasonal drought and frequent fires that evolved in this subtropical climate have served our modern shrub species well. A number of species of chaparral shrubs evolved within tropical families that are otherwise uncommon in our flora.

Scrub Oaks

Typical deeply rooted chaparral shrubs are well exemplified by our scrub oaks. To someone from the eastern United States, these do not look much like oaks, with their shrubby growth form and small, leathery evergreen leaves. The presence of acorns, however, or at least the remnant of acorn caps when the acorns themselves are not present, is a clear indication that these are oaks. Surprisingly, nearly half of the oak species present in California are shrubs. The three common scrub oaks of Southern California are all similar in appearance, and only in recent years have scientists become confident of the limits of their differences.

The common scrub oak through most of our chaparral is *Quercus berberidifolia,* formerly known in many books as *Q. dumosa* (pl. 86). This scrub oak is a common codominant of chaparral stands on steep north-facing slopes, generally above about 800 feet in elevation, and at times may form virtually pure stands on such slopes. Scrub oak typically resprouts rapidly from woody root crowns after fires, forming impenetrable thickets. Individual shrubs reach anywhere from three to 12 feet or more in height depending on growing conditions, but they show multiple trunks with a shrubby architecture. The shiny green leaves are typically small (half-an-inch to an inch in length), flat (not cupped), and sharply spine-toothed along the margins.

Similar at first glance to scrub oak is Muller's scrub oak *(Q. cornelius-mulleri),* which grows in dry chaparral above

Plate 86. Scrub oak (*Quercus berberidifolia,* Fagaceae), March to April.

about 3,000 feet. It occurs in the transition from shrubland to desert in the San Bernardino Mountains and southward along the crest of the Peninsular Ranges from Orange County through San Diego County. Although the small spiny-margined leaves are superficially similar to those of scrub oak, they can be distinguished by a dense cover of fine hairs on the lower leaf surface (pl. 87).

Plate 87. Muller's scrub oak (*Quercus cornelius-mulleri*, Fagaceae), March to April.

The third common scrub oak of Southern California is Tucker oak *(Q. john-tuckeri)*, a species occurring in the upper and more arid chaparral stands near the desert margins of the Tehachapi Mountains and the interior side of the San Gabriel Mountains. Tucker oak is generally larger than scrub oak, reaching from six to 20 feet in height. It can be distinguished from scrub oak by leaves that are dull gray green on the upper surface rather than shiny green, and by thin walls on the acorn cups compared to the thick, warty acorn walls of scrub oak (pl. 88). One place to see Tucker oak is on Hwy. 5 through Tejon Pass in the Tehachapi Mountains. Here, there is an interesting pattern of interbreeding between Tucker oak and

Plate 88. Tucker oak (*Quercus john-tuckeri,* Fagaceae), March to April.

blue oak *(Q. douglasii)* where the ranges of these two species overlap. Blue oak, a deciduous tree, is dominant on less arid north-facing slopes (see pl. 188), whereas the evergreen shrubs of Tucker oak dominate drier south-facing slopes. Slopes with intermediate exposures support oak populations with the unusual blending of growth forms and deciduousness that one might expect from hybrids of such different parents.

Manzanitas and Their Relatives

The common name manzanita comes from the Spanish word *manzana* (apple); the small red fruits of manzanita look like little apples. The manzanita genus, *Arctostaphylos,* is an important chaparral group in California, with 56 species. Fortunately for field identifications in our area, only a few of these are common. A number of other species in Southern California occur in the chaparral of the Channel Islands, as discussed in chapter 10, or at higher elevations in the montane coniferous forests. Along the central and northern coasts of California, where there are large numbers of narrowly distributed species of manzanita, it is more difficult to identify species.

All of our species of manzanita have the evergreen leathery leaves typical of chaparral plants. They hold their alternate leaves relatively vertically, so that it is difficult to tell which is the upper and which is the lower surface. Most species also have distinctive smooth red bark on crooked woody stems.

The responses of manzanita species to fire are important in their ecology, as discussed in more detail in the next chapter, and serve as one clue in their identification. One group of species responds to fire by resprouting from a woody burl or root crown lying just below the soil surface. A careful examination of the base of a manzanita shrub usually reveals whether or not such a burl is present. The other group of species lacks this woody burl and is entirely killed by chaparral fires. This group, however, maintains large numbers of long-lived seeds buried in the soil (called soil seed banks) that are stimulated by the heat of the fire to germinate and recolonize the area.

Far and away the most common of our manzanitas is the Eastwood manzanita *(Arcostaphylos glandulosa)*, a shrub four to eight feet in height that is widespread on rocky slopes and along ridgelines where soils are shallow (pl. 89). It is the only common Southern California chaparral species of manzanita with a basal burl. Paired with this species in many areas is the larger bigberry manzanita *(A. glauca)*, a tall, almost treelike species reaching 15 to 20 feet in height (pl. 90). Bigberry manzanita lacks a burl but produces copious quantities of brownish red fruits. These fruits reach a half-inch in diameter, twice the size typical for manzanita, and give rise to the common name.

Two other shrub species of manzanita are common only at the upper margin of our foothill zone and are not pictured here. Mexican manzanita *(A. pungens)* is found from about 2,500 to 7,000 feet in upper chaparral elevations or in drier habitats in the lower pine forests, particularly in the Peninsu-

Plate 89. Eastwood manzanita (*Arctostaphylos glandulosa* var. *zacaensis*, Ericaceae), January to April.

Plate 90. Bigberry manzanita (*Arctostaphylos glauca*, Ericaceae), January to March.

lar Ranges. The sharp tip of the leaf apex, the source of its species name, and its smaller fruits separate Mexican manzanita from bigberry manzanita, a more common and lower-elevation species. Cuyamaca manzanita *(A. pringlei)* is a montane species more typical of open pine forests and higher-elevation stands of chaparral in the Peninsular Ranges. Both of these species lack a basal burl and instead use soil seed banks to reestablish themselves after fire.

Three additional chaparral shrubs are closely related to manzanita but are each in a different genus. These can be confused with manzanita at first glance and indeed were once considered by scientists to be species of *Arctostaphylos*. The most frequently encountered of these is mission manzanita, which is widespread in coastal chamise chaparral stands of

Plate 91. Palo blanco (*Ornithostaphylos oppositifolia*, Ericaceae), March to April.

Plate 92. Summer holly (*Comarostaphylis diversifolia* var. *planifolia*, Ericaceae), April to May.

San Diego County. One easy way to distinguish this from true manzanitas is to look for a leaf margin that is rolled under, or revolute, in this species (see pl. 82), compared to the flat leaves of manzanita. Less common are two other relatives: palo blanco *(Ornithostaphylos oppositifolia)*, with linear opposite leaves (pl. 91), and summer holly *(Comarostaphylis diversifolia)*, which has warty fruits rather than the smooth berries of manzanita (pl. 92). These latter two species are infrequently encountered in coastal chaparral, particularly in San Diego County and northwestern Baja California.

Ceanothus and Other Nitrogen Fixers

Nitrogen, a critical building block for proteins, is commonly the most important limiting element for growth in chaparral soils. Almost all well-watered chaparral shrubs increase their rates of growth if they are fertilized. Nitrogen, however, is only abundant in the atmosphere, where flowering plants cannot directly utilize it. Several groups of microorganisms have evolved means to fix this nitrogen directly from the atmosphere into their tissues. Most of these microorganisms live as single cells in the soil and help increase soil fertility. Over millions of years, however, a few flowering plants have developed symbiotic relationships with some of these microorganisms to the benefit of both. Where these symbioses exist, the microorganisms form colonies within nodules along the roots of the host plant. These colonies fix nitrogen from the atmosphere and release it into the roots of the plant, providing a source of this limiting element. In return, the plant provides a home and critical carbohydrates and other elements for the healthy growth and reproduction of the bacteria. Although these nitrogen-fixing root nodules occur in relatively few plant groups, a surprisingly large number of these can be found in our chaparral.

The most important nitrogen-fixing shrubs in chaparral are the numerous species of *Ceanothus*. This genus is widespread throughout Southern California; there are at least 52 species of ceanothus in the world, and 43 of these occur in California. Thirteen species can be found in the mainland chaparral of Southern California, with more on the Channel Islands and in the coniferous forests at higher elevations. All of these species have relatively high levels of nitrogen in their leaves and are thus important browse plants for a diversity of wildlife species.

Two major lineages of the genus *Ceanothus* can generally be separated on the basis of a few simple morphological features. The subgenus *Ceanothus* is characterized by thin leaves that have three main veins and are arrayed alternately on the stems, and by the absence of distinctive horns on the fruits. The second and larger group is the subgenus *Cerastes*. These species typically have smaller, leathery leaves with a single main vein and pairs of persistent brownish stipules at the base. The leaves are generally opposite in arrangement on the stems, although this character may be variable in a few species, as described below. Their fruits have three distinctive horns on the surface.

Both *Ceanothus* and *Cerastes* are well represented in chaparral, and indeed it is common to find one species from each group growing in the same elevational zone of an area of foothills. A number of general traits typically separate the groups ecologically. The *Ceanothus* group, with its thin leaves and deeper taproots, is best adapted for survival on soils of north-facing slopes where moisture is more available. The *Cerastes* group has relatively shallow roots and thick drought-adapted leaves, and it has evolved the ability to tolerate shallow and dry soils, such as those present on rocky or south-facing slopes. Most species in both groups are entirely killed by fires but reproduce successfully through soil seed banks that are stimulated to germinate by the heat of a fire.

Four species of the *Ceanothus* subgenus are commonly encountered in our region. From Santa Barbara southward to Orange County the most abundant of these is greenbark ceanothus *(C. spinosus)*. This is a shrubby tree reaching up to 20 feet in height that prefers the deeper soils of north-facing slopes. It is easily recognized by its smooth green bark, flexible branches that end in spiny tips, and leaves with only a single main vein, an exception for this group (pl. 93). Unlike the majority of ceanothus species, greenbark ceanothus has a distinct basal burl and resprouts readily after fires.

Two species occur at higher elevations above about 1,500 feet in the chaparral: hairy-leaved ceanothus *(C. oliganthus)* in the Santa Monica Mountains and white thorn *(C. leucodermis)* in the Transverse and Peninsular Ranges. Hairy-leaved

Plate 93. Greenbark ceanothus (*Ceanothus spinosus*, Rhamnaceae), February to May.

ceanothus has a shrubby tree form of growth like that of greenbark ceanothus, but it can be distinguished by the absence of green spiny branches, the presence of three distinct major veins on the leaf, and downy hairs on the lower leaf surface (pl. 94). White thorn is a shrubby species with rigid spiny branches and a powdery white coating on the upper surface of the leaves (pl. 95).

Although both greenbark ceanothus and hairy-leaved ceanothus are present in San Diego County, they are less com-

Plate 94. Hairy-leaved ceanothus (*Ceanothus oliganthus*, Rhamnaceae), March to April.

Plate 95. White thorn (*Ceanothus leucodermis*, Rhamnaceae), April to May.

Plate 96. Woolly-leaf ceanothus (*Ceanothus tomentosus*, Rhamnaceae), April to May.

mon and are largely replaced along an elevational gradient by other species. In the lower elevations of coastal chaparral the dominant species is woolly-leaf ceanothus *(C. tomentosus)*, a shrub with dense white hairs on the lower leaf surface (pl. 96). Higher-elevation chaparral stands away from the coast are characterized by white thorn, although hairy-leaved ceanothus remains widespread.

We can nicely pair four species of the *Cerastes* group with the above *Ceanothus* species. The dominant lower-elevation

Plate 97. Bigpod ceanothus (*Ceanothus megacarpus,* Rhamnaceae), January to April.

species from Santa Barbara County southward through Orange County is bigpod ceanothus *(Ceanothus megacarpus),* which has small leathery leaves (pl. 97). This shrub germinates in massive numbers after fires in the southern coastal ranges and Santa Monica Mountains, forming almost pure stands on more arid south-facing or rocky slopes. Many south-facing chaparral slopes turn white in January with flowering masses of this shrub, the first abundant species to bloom. At higher elevations bigpod ceanothus is replaced by hoary-leaved ceanothus *(C. crassifolius),* a more open-branched large shrub with small leathery leaves that are olive green above but have dense, woolly white hairs below (pl. 98).

Like greenbark ceanothus, these two species become less important in San Diego County, where they are replaced at low elevations by wart-stemmed ceanothus *(C. verrucosus)* and at higher elevations by cup-leaf ceanothus *(C. greggii).* The former species has distinctive stem morphology with warty growths (stipules) and irregularly angled stems, and

Plate 98. Hoary-leaved ceanothus (*Ceanothus crassifolius*, Rhamnaceae), March to April.

Plate 99. Cup-leaf ceanothus (*Ceanothus greggii* var. *perplexans*, Rhamnaceae), March to April.

unlike typical *Cerastes* species has a major portion of its leaves alternate in arrangement (see pl. 81). Cup-leaf ceanothus lacks these warts along its rounded stems and generally exhibits cup-shaped leaves (pl. 99).

Other uncommon or rare species of ceanothus may be encountered in the chaparral of Southern California but most of these are not illustrated here. Three are members of the subgenus *Ceanothus.* Deerbrush *(Ceanothus integerrimus)* is a widespread higher-elevation species that extends into the lower montane zone of pine forests. It has spiny branches and loses its leaves in winter. Palmer ceanothus *(C. palmeri)* is another deciduous species of higher elevations that is most common from Mount San Jacinto southward along the inner Peninsular Ranges of San Diego County. Lakeside ceanothus *(C. cyaneus)* is a rare species occurring at low elevations in the same ranges of San Diego County (pl. 100). Among the *Cerastes* species, wedgeleaf ceanothus *(C. cuneatus)* is occasionally found at higher elevations of chaparral throughout our region, but it is more typical of the foothills of the Sierra Nevada and the inner slopes of the Coast Ranges.

Plate 100. Lakeside ceanothus (*Ceanothus cyaneus,* Rhamnaceae), April to June.

Although ceanothus species are the most widespread group of nitrogen-fixing shrubs in the chaparral, they are not the only ones. Several shrub species in the rose family have also developed nitrogen-fixing symbioses. The most common of these are the mountain mahoganies, with two species in our area. California mountain mahogany *(Cercocarpus betuloides)* is a widespread species that ranges well beyond the chaparral of Southern California to many other communities in the western United States. In chaparral it is typically a branched shrub six to 15 feet or more in height, but in some mixed woodland communities it may take on the form of a small tree. The distinctive leaves of this species are cupped, have leaf margins that are smooth in the basal half of the leaf but toothed above, and have whitish downy hairs on their lower surface (pl. 101). The unique fruits have a silky plume two to three inches in length to aid in wind dispersal. Wind dispersal is quite unusual among chaparral shrubs, which typically have fruits that are dispersed by birds or mammals. The common name mountain mahogany comes from the dense nature of the woody tissue of the stems. The Native Americans in our area used these as digging sticks and for spear and arrow tips. A second and less common species is San Diego mountain mahogany *(C. minutiflorus)*, which occurs in San Diego County and northwestern Baja California. It is greenish yellow on its lower leaf surface and lacks hairs. As its scientific name suggests, it has very small flowers compared to its more common relative.

Plate 101. California mountain mahogany (*Cercocarpus betuloides*, Rosaceae), March to June.

An unusual nitrogen-fixing member of the rose family is southern mountain misery *(Chamaebatia australis)*, a shrub with highly dissected and sticky leaves (pl. 102). This species is largely restricted to scattered locations in San Diego County and adjacent Baja California. It is closely related to a much more common species of mountain misery in the Sierra Nevada foothills at the lower margin of the ponderosa pine forests. The unusual common name of this species comes from the easy transfer of the pungent odor of the sticky foliage to shoes and pants.

Plate 102. Southern mountain misery (*Chamaebatia australis,*
Rosaceae), April to May.

The final group of nitrogen-fixing shrubs comprises
members of the legume or pea family (Fabaceae) that have
their own special forms of bacterial symbiosis. Many herbaceous legumes, such as lupines and lotus species, are associated with chaparral stands after fire and with grasslands.
These are discussed in chapters 5 and 7. Only one legume
species forms a truly woody chaparral shrub in our region.
This is chaparral pea *(Pickeringia montana)*, a stiff woody
shrub with spiny branches that often reach to six to eight feet
in height (pl. 103). The small evergreen leaves of chaparral
pea are formed of three leaflets.

Plate 103.
Chaparral pea
(*Pickeringia
montana,*
Fabaceae),
April to May.

Some Tropical Origins

There are two groups of chaparral shrubs that do not readily fit the categories above. These come from three families: the cashew family (Anacardiaceae), the silk tassel family (Garryaceae), and the cacao family (Sterculiaceae). Each has direct evolutionary linkages to tropical ancestors. One way to adapt to summer-dry conditions is to develop deep roots that can tap underground water resources. Although all chaparral shrubs have relatively deep roots to get them through summer drought periods, several species in the family Anacardiaceae are particularly notable for the way that they can tap into deep groundwater sources. The most exceptional of these is laurel sumac *(Malosma laurina)*, a common and widespread evergreen shrub that reaches 10 to 15 feet in height in lower-elevation chaparral and the margins of oak woodland communities of Southern California (pl. 104). Laurel sumac has deep roots that allow it to virtually ignore summer drought conditions and continue growth throughout most of the year.

Plate 104. Laurel sumac (*Malosma laurina,* Anacardiaceae), June to July.

Laurel sumac is also able to thrive in relatively dry coastal sites as long as fractured bedrock allows it to send its roots down deeply into the soil. This species is sensitive to even light frosts, however, as ice crystals form easily in its water-conducting tissues. Early ranchers in Southern California learned this trait and used laurel sumac as an indicator for citrus planting. If a site was too cold for laurel sumac, frost would also damage citrus crops there. Although frosts can readily

Plate 105. Sugar bush (*Rhus ovata,* Anacardiaceae), March to May.

kill back the stems of laurel sumac, the large root crown of this species allows it to resprout rapidly following damage from frost or fire.

Laurel sumac has two close relatives in the chaparral of Southern California. These are lemonadeberry *(Rhus integrifolia)*, discussed in chapter 3, and sugar bush *(R. ovata)*. Both are evergreen shrubs five to 10 feet in height with thick main trunks and glossy green leaves. The two commonly separate ecologically, however, with lemonadeberry occurring at lower elevations on the ocean-facing side of the coastal ranges, and sugar bush typically growing away from the coast at elevations above 1,000 feet. The thick ovate leaves of sugar bush are about two-and-a-half inches in length and folded down the middle in what has been described as a tacolike shape (pl. 105). The leaves of lemonadeberry, in contrast, are typically slightly smaller and flat (see pl. 35).

The silk tassels of the genus *Garrya* (family Garryaceae) are another subtropical shrub group with existing close relatives in Central America and the Caribbean. Three of the six silk tassel species in our state can be found in the chaparral of

Southern California. The most common of these is southern silk tassel *(Garrya veatchii)*, an evergreen shrub reaching to 10 feet in height and occurring on drier chaparral slopes, often at higher elevations. At first glance its vertically oriented leaves with similar upper and lower surfaces suggest a manzanita (pl. 106). Silk tassel bushes always have opposite leaves, however, whereas manzanita leaves are alternately arranged. A second species not pictured here, ashy silk tassel *(G. flavescens)*, occurs in chaparral along the edge of the desert and occasionally enters dry chaparral communities at higher elevations.

An unusual pair of chaparral shrubs in the cacao family (Sterculiaceae), a typically tropical family, are the flannelbushes. California flannelbush *(Fremontodendron californicum)* is a tall chaparral shrub up to 12 feet or more in height that is common in higher-elevation chaparral stands on the desert margins of the Transverse and Peninsular Ranges. It has distinctive palmately lobed leaves and large lemon yellow flowers up to 3 to 4 inches across (pl. 107). Similar in appearance is Mexican flannelbush *(F. mexicanum)*, a shrub with a

Plate 106. Southern silk tassel *(Garrya veatchii,* Garryaceae), February to April.

Plate 107. California flannelbush (*Fremontodendron californicum*, Sterculiaceae), April to June.

Plate 108. Mexican flannelbush (*Fremontodendron mexicanum,* Sterculiaceae), April to June.

more erect growth form that is unbranched near the base (pl. 108). It is restricted to scattered chaparral areas of Orange, San Diego, and Imperial Counties as well as adjacent areas of northwestern Baja California.

Deciduous Chaparral Shrubs

Having said that chaparral shrubs are typically evergreen, we need to acknowledge a few successful species that do not follow this rule. Surprisingly, these deciduous chaparral shrubs lose their leaves in winter, not in summer. This pattern contrasts sharply with that of the drought-deciduous shrubs that characterize coastal sage scrub communities, which lose their leaves with the onset of summer water stress. Winter leaf loss, a characteristic shared with our deciduous oaks and many riparian tree species, such as sycamores (*Platanus* spp.) and alders (*Alnus* spp.) (see chapter 8), undoubtedly derives from genetic relationships with ancestral species that grew in regions with more pronounced winter cold. There appears to be some gradual evolution away from this winter-deciduous pattern, however, in a few groups. Gooseberries and currants (*Ribes* spp.), as well as California buckeye *(Aesculus californica)* in blue oak woodlands (see chapter 6), lose their leaves early and flush new ones in midwinter.

The most commonly encountered deciduous chaparral shrubs are species of gooseberries and currants. Both are members of the genus *Ribes;* gooseberries have spines, currants do not. As many as five species may be encountered in the chaparral of Southern California, but only three of these are common. Two other species are discussed in chapter 6 as components of oak woodland communities. White-flowered currant *(Ribes indecorum),* sometimes called white chaparral currant, is an erect shrub three to six feet in height that is common in chaparral and coastal sage scrub over our area (pl. 109). Its leaves are one to one-and-a-half inches wide. Chaparral currant *(R. malvaceum)* is similar in appearance vegetatively but has leaves one-and-a-half to two inches wide (pl. 110) and pink or rose-colored flower clusters. Chaparral currant favors somewhat wetter chaparral sites and can be found in oak woodland communities as well.

The third *Ribes* species commonly encountered in chaparral as well as coastal sage scrub and open woodlands is the fuchsia-flowered gooseberry *(R. speciosum).* As its designation as a gooseberry indicates, this species is highly spiny. Stems with three spines at each leaf node and small oval leaves

Plate 109. White-flowered currant (*Ribes indecorum*, Grossulariaceae), October to March.

Plate 110. Chaparral currant (*Ribes malvaceum*, Grossulariaceae), October to March.

Plate 111. Fuchsia-flowered gooseberry (*Ribes speciosum,* Grossulariaceae), March to May.

that are shiny green on their upper surface make this an easy species to recognize (pl. 111). The common name comes from the resemblance of the small pendulous red flowers to those of fuchsias. Flowering begins as early as January, well before that of most other shrub species, and provides an important food source for hummingbirds that pass through Southern California on their annual paths of migration. Indeed, early winter flowering is a characteristic of all of our currants and gooseberries.

The final deciduous woody shrub found widely is chaparral ash *(Fraxinus dipetala),* also called foothill ash or California ash. We normally think of ashes as tall trees, but this species and another in northwestern Baja California have evolved into shrubby growth forms. Perhaps because its close relatives are trees, chaparral ash retains a single main trunk. It commonly reaches only six to eight feet in height, although on more favorable sites it can become treelike and reach 20 feet. Like most species of ash in California, chaparral ash has compound leaves, oppositely arrayed on its branches and di-

Plate 112. Chaparral ash (*Fraxinus dipetala*, Oleaceae), April to May.

vided into three to seven ovate leaflets with finely toothed margins (pl. 112). Another distinctive feature of chaparral ash is that the young branches are distinctly four angled. Chaparral ash is one of the rare chaparral shrubs that has seeds dispersed by the wind rather than by animals. The membranous wing extension on the seeds allows them to float like gyrocopters for long distances in the wind.

Showy Herbaceous Perennials

Although mixed chaparral communities are dominated by large woody shrubs whose flowers are individually small, a number of tall herbaceous perennials with showy flowers can be found in openings between shrubs throughout our region. Among the most spectacular of these are the penstemons, with their large, colorful flowers and opposite leaves clasping the stem. Scarlet bugler *(Penstemon centranthifolius)* displays an array of tubular scarlet flowers and elongate but not linear leaves that strongly clasp the stem (pl. 113); the slender

Plate 113. Scarlet bugler (*Penstemon centranthifolius*, Scrophulariaceae), April to May.

Plate 114. Foothill penstemon (*Penstemon heterophyllus* var. *australis*, Scrophulariaceae), April to June.

foothill penstemon *(P. heterophyllus)* displays multiple stems of tubular rose violet flowers and linear leaves (pl. 114). Even more dramatic is showy penstemon *(P. spectabilis)*, which has spikes of large rose purple flowers with pale centers (pl. 115). Its leaves are broader than those of the other species and distinctly toothed along their margins.

Plate 115. Showy penstemon (*Penstemon spectabilis,* Scrophulariaceae), April to May.

Heart-leaved penstemon *(Keckiella cordifolia)* is a sprawling semiwoody shrub closely related to the penstemons that colonizes disturbed areas on semishaded slopes and clambers over its neighbors (pl. 116). It can also be found in oak woodlands and coastal sage scrub. Its common name comes from its small heart-shaped leaves, which are arranged oppositely on the stems. It has sometimes been called scarlet honeysuckle because its tubular red flowers resemble those of the true honeysuckles (*Lonicera* spp.). A related species with yellow flowers, yellow bush penstemon *(K. antirrhinoides),* can be found in the arid interior chaparral of the Peninsular Ranges (pl. 117).

Several other plants that colonize open areas are members of the legume family. One of these is bush lupine *(Lupinus longifolius),* a shrubby species that can reach four to five feet

Plate 116. Heart-leaved penstemon (*Keckiella cordifolia,* Scrophulariaceae), March to August.

Plate 117. Yellow bush penstemon (*Keckiella antirrhinoides,* Scrophulariaceae), April to June.

in height. It has typical lupine foliage with palmately divided leaflets and the distinctive characteristic of silky hairs covering both sides of the leaflets (pl. 118). Several trailing herbaceous legumes that use tendrils to climb over shrubs are common along trails and roads in chaparral and live oak woodlands. Common among these are two species of wild sweet pea, canyon pea *(Lathyrus vestitus)* and pride of California *(L. splendens).* The former is widespread throughout California, whereas the latter is restricted to more open habitats of San Diego County and adjacent Baja California (pls. 119, 120).

Perhaps the showiest large herbaceous perennial growing in mixed chaparral is matilija poppy, which is almost shrubby in that it has a distinctly woody base. Its tall leafy stems typically reach six to eight feet in height and extend creeping rhizomes (underground stems) outward to form dense clumps of growth. Matilija poppy has the distinction of having the largest flower of any species in California. The broad "fried egg" flowers are yellow in the center and white on the petals,

Plate 118. Bush lupine (*Lupinus longifolius*, Fabaceae), April to June.

Plate 119.
Canyon pea
(*Lathyrus
vestitus*,
Fabaceae),
March to May.

Plate 120. Pride of California (*Lathyrus splendens*, Fabaceae), April to June.

with a width of up to four inches (pl. 121). Our region is home to two very similar species, the canyon matilija poppy *(Romneya coulteri)* and the coastal or hairy matilija poppy *(R. trichocalyx)*. The latter species is distinguished by hairs on its sepals.

Plate 121. Canyon matilija poppy (*Romneya coulteri*, Papaveraceae), May to July.

Mixed Chaparral–Oak Woodland Transitions

Several large chaparral shrub species commonly occur in transition zones at the margins of oak woodland (pl. 122). These species, unlike the majority of chaparral shrubs, are relatively shade tolerant and thus can survive under the canopies of larger live oaks or other woodland species. The most common of these shrubs is toyon *(Heteromeles arbutifolia)*. Although toyon is typically a large shrub six to eight feet in height, it frequently takes on a treelike growth form and may reach heights of 25 feet or more (pl. 123). Toyon produces abundant crops of red berries in winter, and thus Americans settling a century ago in Southern California called it California holly. It was the abundance of this species in the hills above Los Angeles that gave rise to the name Hollywood. The fruits are sweet and attract a myriad of birds in winter and early spring. Native Americans collected these berries as well and boiled them as an important food, and early Spanish settlers used them to brew a pleasant beverage.

Plate 122. Transitional habitat between mixed chaparral and live oak woodlands on north-facing slopes in the Santa Monica Mountains.

Plate 123. Toyon (*Heteromeles arbutifolia,* Rosaceae), June to July.

Another member of the rose family that favors richer chaparral habitats is hollyleaf cherry *(Prunus ilicifolia).* Like toyon, this is an evergreen shrub that can become treelike under optimal growing conditions; it can reach 30 feet in height. Its lustrous evergreen leaves and profusion of small white flowers in spring are beautiful (pl. 124), and thus hollyleaf cherry has been widely used in horticulture. The leaves are ovate in general shape but sharply toothed along their margins and curled such that they would not lie flat on a horizontal surface. Unlike the related domestic cherries and other stonecrop fruits that are highly edible, the fruits of this species have only a thin sweet flesh over a large seed. Nevertheless, Native Americans are reported to have brewed these fruits into an alcoholic beverage.

An easy shrub species to confuse at first sight with hollyleaf cherry if the two are seen separately is hollyleaf redberry *(Rhamnus ilicifolia).* As the species names indicate, both have hollylike toothed leaves, and both are shrubs or shrubby trees occurring in similar habitats of richer soils adjacent to or intermixed with coast live oaks. Although the leaves are similar in size, hollyleaf redberry leaves lack the lustrous green surface of hollyleaf cherry leaves, are less sharply toothed, and are flattened so as to lie in a single plane (pl. 125). In early summer, when fruits are mature, the two species can be easily distinguished. Hollyleaf redberry has

Plate 124. Hollyleaf cherry (*Prunus ilicifolia*, Rosaceae), March to May.

small fleshy red berries, whereas hollyleaf cherry has much larger dark purple fruits (a half-inch or more in diameter) with a thin flesh over a large seed.

Two other species closely related to hollyleaf redberry are

Plate 125. Hollyleaf redberry (*Rhamnus ilicifolia*, Rhamnaceae), March to April.

Plate 126.
Coffeeberry
(*Rhamnus californica*,
Rhamnaceae), May
to June.

found in the chaparral of Southern California. Coffeeberry
(*R. californica*) grows as a mounded to tall shrub six to 15 feet
or more in height. Like the other species in this group, it may
be found infrequently in richer chaparral soils or within oak
woodlands throughout Southern California. The leaves of
coffeeberry are one to three inches in length and have margins
that are slightly curled toward the lower leaf surface, which
lacks hairs (pl. 126). Very similar is hoary coffeeberry (*R. tomentella*), not pictured here, which has thick hairs on the
lower leaf surface. This latter species is most common in
mixed chaparral stands in the Peninsular Ranges of San Diego
County. The common name coffeeberry comes from the resemblance of the red ripe fruits to coffee beans. Early European settlers were disappointed to find that the similarity did
not extend to taste. Native Americans and early European settlers used the bark of both species to make a laxative.

Baja Mixed Chaparral

A distinctive form of mixed chaparral can be found on several
coastal peaks in southern San Diego County adjacent to the
Mexican border in areas underlain by an unusual dark volcanic rock that weathers to form soils inhospitable to many
chaparral shrubs (pl. 127). The resulting chaparral commu-

Plate 127. Baja mixed chaparral community on Otay Mountain in San Diego County, with manzanita, bush poppy, fragrant sage, and southern mountain misery.

Plate 128. Tecate cypress (*Cupressus forbesii*, Cupressaceae).

nity represents the northern limit of a form more typical of northwestern Baja California. The dominant shrub cover of this Baja mixed chaparral includes several species illustrated elsewhere in this book. These are southern mountain misery (see pl. 102), woolly-leaf ceanothus (see pl. 96), and chamise (see pl. 73). It is here that one finds the endemic Otay manzanita *(Arctostaphylos otayensis)*, as well as local populations of Tecate cypress *(Cupressus forbesii)*, a rare shrubby conifer 20 to 30 feet in height (pl. 128). A small, isolated stand of Tecate cypress also occurs in the foothills of the Santa Ana Mountains in Orange County.

Fire as a Natural Component of Chaparral Environments

Anyone living in California is well aware of the manner in which intense fires can burn rapidly through the foothills. Dry fall conditions and Santa Ana winds can combine to produce intense chaparral fires that move rapidly across the slopes and canyons of Southern California, often resulting in massive property damage to homes in the foothills (pl. 129). Despite the destruction that such large fires can bring, it is important to remember that fire is a natural component of our chaparral environment. Although most chaparral fires today are caused by humans, either through carelessness or by deliberate arson, natural fires ignited by occasional summer lightning strikes have occurred for thousands of years. The native plants of our chaparral communities exhibit a wide range of remarkable adaptations enabling them either to survive fires or to recover their populations after fires.

Natural lightning fires in chaparral typically occur when thunderstorms originating in the summer rainfall areas to the east move across Arizona and force their way into Southern California. Such storms are rare along our coast and exceedingly rare on the Channel Islands, but are more frequent in the Transverse and Peninsular Ranges and areas to their east. Lightning storms generally peak in August and September, before most of our Santa Ana winds develop. Without these winds and the associated high temperatures and low relative humidities, chaparral fires burn much less intensely.

In contrast, human-caused chaparral fires peak in fall, when chaparral foliage is exceedingly dry following months of summer and early fall drought. Fall is also when strong Santa Ana wind conditions can occur, spreading the flames. Ignition of shrub foliage becomes difficult once winter rains commence and stems and leaves rehydrate. An easy way to think about this seasonal difference in flammability is to con-

Plate 129. An advancing fire at Lake Elsinore, Riverside County (photograph by Jon Keeley).

sider the ease of starting a fire at home in your fireplace with dry kindling (like summer chaparral foliage) compared to green stems (like winter foliage with a high water content). Combine the ease of igniting dry chaparral foliage with strong Santa Ana wind conditions and you have all the ingredients for a massive fire.

It is not easy to determine the natural frequency of chaparral fires. The records of fire occurrences in Southern California go back nearly a century. These data show that some areas of the Santa Monica Mountains, for example, have burned nine times in the past 80 years, while other areas have not burned even once. These differences are likely due to a variety of causes, including human access, with more access meaning more fire ignitions, and the topography of sites in relation to ocean breezes and protection from Santa Ana winds. Before humans came on the scene, a best guess is that fire frequency was probably somewhere between 30 and 100 years, although there is no direct way of telling this analogous to the fire scars on tree rings in conifer forests.

Native Americans, who became widespread in California

beginning about 6,000 to 10,000 years ago, made wide use of fire in managing their environment. They burned chaparral areas to increase the local availability of wildlife, to promote healthy regrowth of willows and other plants used in basketry, and to increase the populations of bulb plants, which they used for food. These early fire management practices changed dramatically at the end of the nineteenth century as Native American dominance declined and new American ideas of resource management focused on fire suppression. A century of fire suppression in chaparral has reduced fire frequency, but chaparral fires are more intense when they do occur because the mass of unburned fuels has increased.

Shrub Response to Fire

Chaparral fires frequently reach temperatures of 1,200 degrees F or more, making them among the hottest fires in any natural environment in the world today. Not all chaparral fires burn with the same intensity, however, and the intensity of any given fire varies over the area burned. Temperature, relative humidity, and wind speed, as well as biological variables such as stand age and foliage and stem moisture content, all influence fire intensity. Even with all of these factors equal, fires burn much more intensely moving up a slope than down. Thus, an area burned in a large chaparral fire is a mosaic of more and less intensely burned slopes.

If you visit a chaparral slope after a fire, you can gauge the relative intensity of the burn by walking through the gray ash and looking at the blackened stems of the former shrubs (pl. 130). The diameter of the smallest remaining branches is directly related to the fire intensity. A very intense fire consumes all of the aboveground woody tissues, leaving only woody nubs to show where shrubs formerly grew. A fire of moderate intensity leaves blackened stalks of the main shrub

Plate 130. Bundy Canyon, in the Santa Monica Mountains, the day after the 1962 fire (photograph by Laurel Woodley).

branches, whereas a low-intensity fire leaves even small stems as thin as a quarter-inch in diameter.

Chaparral shrubs have two primary strategies for postfire recovery. Most species have woody root crowns or burls below the soil surface that are protected from the heat of the fire, and these readily resprout multiple new stems (pls. 131, 132). Because these belowground tissues store carbohydrates, resprouting can occur relatively rapidly, even in the absence

Plate 131. Regeneration of the chaparral in Decker Canyon, Santa Monica Mountains, about one month after fire burned through the area.

Scrub oak

Chamise

Ceanothus

Plate 132. Resprouting of shrubs from underground root crowns or stems shown one month after a fire.

Chaparral yucca and chamise

of rain. Resprouting from underground tissues is considered to be an ancestral evolutionary condition in many shrubs throughout the world and is not necessarily a fire adaptation. Although fire can provide a cue for resprouting, loss of above-ground stems to grazing or frost damage frequently causes the same response.

As mentioned in the previous chapter, some groups of species of manzanita *(Arctostaphylos)* and ceanothus *(Cean-*

Plate 133. Chamise and ceanothus seedlings germinated in the spring after a fire.

othus) lack these root crowns and are entirely killed by chaparral fires. These species deposit long-lived seeds in the soil beneath their canopies. The heat of a fire stimulates these seeds to germinate. Although the parent shrub dies, large numbers of offspring seedlings compete to take its place in the new chaparral canopy (pl. 132). With a few exceptions, chaparral shrub species reestablish after fire either by resprouting or by reseeding, with no intermediates. The major exception is chamise, which is an excellent resprouter but also comes up in hundreds of new seedlings after fires (pl. 133). The reseeding strategy of chamise, however, is very different from that of ceanothus and manzanitas. Chamise seeds are small and relatively short lived in soil seed banks. They must be constantly replenished from the parent shrubs.

Showy Fire-Following Annuals

One of the most spectacular wildflower shows to be seen anywhere in the world is that which occurs in the first spring after

Plate 134. A showy display of fire-following annuals in the first spring after a fire, with yellow monkeyflower, Parry's phacelia, large-flowered phacelia, fire poppy, and eucrypta.

a chaparral fire (pl. 134). Slopes that formerly supported dense and impenetrable stands of chaparral shrubs suddenly erupt in a diverse riot of color as dozens of species of annual plants burst into flower in March and early April. Under these conditions, particularly if there have been good fall and winter rains, as many as 50 to 100 or more annual species may be found flowering on a single slope. Besides providing a show for us, these annuals play an important role in helping to protect vulnerable chaparral slopes from erosion in the first year after a fire, when little regrowth of shrubs has occurred.

Where have all of these annual wildflowers been hiding? Most of the species have been patiently waiting as seeds in the soil, often for decades or more. The cue stimulating these seeds to germinate is typically not the heat of the fire, as it is for the larger seeds of ceanothus and manzanitas, but rather the release of nitrogen compounds from the ash of the fire as organic matter begins to be broken down by microorganisms. The annuals that require this cue to germinate are called *fire followers* because they are only rarely found other than in the first year after a chaparral fire. They grow to maturity and shed their seed that year, and the seeds then sit dormant in the soil until the next fire many years in the future.

Although many of the showiest and most abundant post-fire annuals are fire followers, more generalist annual species are often present as well. These are species whose seeds are readily transported by wind or other means and that can quickly colonize the open habitats available after fire. Most of these species are not restricted to postfire occurrences but may also be present in coastal sage scrub and woodlands, as well as along roadsides or other disturbed sites.

We have chosen some of the showiest and most common of the many postfire chaparral annuals to describe and illustrate here. In looking at this or other books describing the size of annuals here or in oak woodlands and grasslands, keep in mind that the sizes of the plants and flowers are highly dependent on the growing conditions in the spring following the fire. If the spring is moist, these annuals can grow taller than they can if it is dry.

Four herbaceous annuals in the poppy family (Papaveraceae) are prominent on chaparral slopes in the first year after fire and can be considered fire followers. Like all members of the poppy family, these species have flowers with just four petals. Two of them have dramatic, showy brick red flowers. These are the fire poppy *(Papaver californicum)* and the wind poppy *(Stylomecon heterophylla).* Both of these poppies are stimulated to germinate in large numbers by the ash left after a fire but are only rarely encountered if there has not been a fire. Typically, fire poppies reach a foot or more in height; they have finely divided leaves with oblong segments and a slender stem with milky sap (pl. 135). The flowers have petals with a greenish basal spot. The wind poppy has similar foliage, but its stems have a yellowish sap and its flower petals have a brown basal spot (pl. 136).

Plate 135. Fire poppy (*Papaver californicum*, Papaveraceae), April to May.

The other two common postfire poppy species are the familiar California poppy *(Eschscholzia californica)* (pl. 137) and a related species called collarless California poppy *(E. caespitosa)*, a yellow-flowered species lacking the California poppy's distinctive ring of tissue around the base of the petals (pl. 138). California poppy, the state flower of California, is extremely variable in its growth form and ecological preferences. It is common after fires in chaparral but is also frequently encountered in grasslands, coastal sage scrub, and other open habitats. The beauty and charm of this species have led to its widespread introduction in horticulture. So adaptable is it that it has escaped to become a widespread and abundant weed along roadsides and disturbed ground in the mediterranean climate region of central Chile.

Plate 136. Wind poppy (*Stylomecon heterophylla*, Papaveraceae), April to May.

Plate 137. California poppy (*Eschscholzia californica*, Papaveraceae), April to May.

Plate 138. Collarless California poppy (*Eschscholzia caespitosa*, Papaveraceae), April to June.

Native Californians cooked the herbage of California poppies as a vegetable, and the early Spanish settlers fried the plants in olive oil and added perfume to make a hair gel.

Another distinctive group of postfire annuals in California chaparral is the phacelia family (Hydrophyllaceae). This family includes many relatively strict fire followers that are only rarely present in the absence of fire, as well as generalist annuals occurring in a variety of plant communities from the desert to oak woodlands. Members of the phacelia family have flowers distributed along a coiled stem, with five petals united at their base or over much of their length. The stems of most species are covered with bristly or sticky hairs.

The showiest and most characteristic members of this family, and indeed the classic fire-following chaparral annuals, are species of the genus *Phacelia,* a large group with nearly 100 species in California. Large-flowered phacelia *(P. grandiflora)* is the largest and often the most abundant of the phacelias that occur widely after chaparral fires (pl. 139, *top left*). Although a relatively short lived annual, it can reach two to three feet in height. As the species name indicates, the bluish to lavender, saucer-shaped flowers are large, with a diameter of one to two inches. The uninitiated often rush into the midst of these extremely attractive flowers to pick a bouquet. More experienced hikers generally suggest some caution in this, however. The sticky hairs of many phacelias, including this one, contain a chemical that readily stains hands and clothing a rusty red brown. A small but significant percentage of people also react badly if their skin comes in contact with this chemical and develop a contact dermatitis resembling that caused by poison oak.

Sticky phacelia *(P. viscida)* is very similar but much smaller in stature. Flowering stems reach one to two feet in height and are topped with coils of flowers similar in form to those of large-flowered phacelia but only about three-quarters of an inch across (pl. 139, *top right*). The basal portions of the petals are distinctly white. Like its larger relative, it has glandular hairs along its stems that quickly stain hands and clothing. Indeed, sitting on either of these species will almost certainly leave a permanent record on the back of one's pants! Although commonly encountered after fires in the northern portions of our region, sticky phacelia is relatively rare in San Diego County.

Parry's phacelia *(P. parryi)* (pl. 139, *middle left*) and California bells *(P. minor)* (pl. 139, *middle right*) appear superficially similar to sticky phacelia but can be distinguished by

Large-flowered phacelia (*Phacelia grandi-flora*, Hydrophyllaceae), February to June.

Sticky Phacelia (*Phacelia viscida*, Hydrophyllaceae), March to May.

Parry's Phacelia (*Phacelia parryi*, Hydrophyllaceae), March to May.

California bells (*Phacelia minor*, Hydrophyllaceae), March to June.

Yellow-throated phacelia (*Phacelia brachyloba*, Hydrophyllaceae), May to June.

Fern-leaf phacelia (*Phacelia distans*, Hydrophyllaceae), March to June.

Plate 139. Common postfire phacelias.

several characteristics. Their flowers are distinctly royal purple, rather than the bluish or lavender of sticky and large-flowered phacelia, and typically have longer, more tubular bases that can be described as bell shaped. This flower form is particularly evident in California bells, whose petals are joined for well over half of their length to form an elegant floral tube. Parry's phacelia has a more open floral tube that lacks this constriction. Both species have long stamens that extend well beyond the length of the petals. Although glandular hairs are present, these species lack the extreme stickiness of large-flowered and sticky phacelia.

Two more common fire followers among the phacelias are easily separated from the four species described above. Yellow-throated phacelia *(P. brachyloba)* is a white-flowered species made distinctive by a prominent yellow throat at the base of the petals (pl. 139, *bottom left*). It is commonly multi-branched, reaches six to 24 inches in height, and has stems covered by thin downy hairs. Unlike all of the species described above, which have moderately ovate and unlobed leaves, the yellow-throated phacelia has basal leaves that are deeply lobed almost to the midrib. Fern-leaf phacelia *(P. distans)* takes this lobing one level further, forming highly dissected leaves that resemble those of a fern. The half-inch to three-quarter-inch blue violet flowers have an open bell shape, with short stamens that do not exceed the length of the petals (pl. 139, *bottom right*).

The less attractive caterpillar phacelia *(P. cicutaria),* a relatively late flowering species, is easily recognized by its long bristly hairs and pale blue to lavender flowers (pl. 140). Although common on chaparral burns, caterpillar phacelia is a weedy, opportunistic species that readily colonizes roadsides, trails, and disturbed sites in live oak woodlands.

Another of the most prominent and abundant postfire annuals in chaparral is a close relative of the phacelias. This is whispering bells *(Emmenanthe penduliflora),* a small annual reaching no more than 12 inches in height. The cream-

Plate 140. Caterpillar phacelia (*Phacelia cicutaria* var. *hispida*, Hydrophyllaceae), March to June.

colored flowers of whispering bells have petals divided almost to their base to produce an open bell shape (pl. 141). The common name comes from these bell-shaped flowers and from the rustling noise made in the wind by the dry hanging flower stalks in late spring and summer. The leaves are linear to oblong and deeply lobed almost to their midrib. Distinctive races of whispering bells occur in the Mojave and Sonoran

Plate 141. Whispering bells (*Emmenanthe penduliflora*, Hydrophyllaceae), April to May.

Deserts and in serpentine grasslands of California, although neither of these latter habitats experiences natural fires.

A third important group of fire-following annuals is the figwort family (Scrophulariaceae). Prominent among these are native snapdragons. The two common species in our area are white snapdragon *(Antirrhinum coulterianum)* and twining snapdragon *(A. kelloggii).* Both are tall annuals reaching up to four feet in height. They have typical snapdragon flowers, which require large insect pollinators to open the closed flower mouths. The two-lipped flowers have three joined petals on the lower half forming a sac, and two smaller petals above forming wings. White snapdragon has white flowers tinged with purple arrayed densely along a main stem for up to 12 inches (pl. 142). The lower flowers mature first; development moves slowly upward such that the lower flowers are already in seed by the time the upper flowers reach maturity.

Plate 142. White snapdragon (*Antirrhinum coulterianum,* Scrophulariaceae), April to June.

Twining snapdragon is less apparent at a distance because of its growth habits but is easily distinguished. Each of the long flower stems of this species begins at the base of a leaf. The stems, two to four-and-a-half feet in length, twine through

Plate 143. Twining snapdragon (*Antirrhinum kelloggii*, Scrophulariaceae), March to May.

Plate 144. Blue toadflax (*Linaria canadensis*, Scrophulariaceae), March to April.

adjacent shrubs to gain support. The flowers are blue to violet in color (pl. 143).

Another tall, slender postfire annual in the figwort family is blue toadflax *(Linaria canadensis)*. This sparsely leaved species reaches up to two feet in height, with short spreading branches only at its base. The two-lipped flowers differ significantly from those of the snapdragons, having a long downward-pointing spur at the base of the floral tube (pl. 144). At times the fragrant blue violet flowers of this species

Plate 145. Yellow monkeyflower (*Mimulus brevipes,* Scrophulariaceae), April to May.

Plate 146. Scarlet larkspur (*Delphinium cardinale,* Ranunculaceae), May to June.

can cover huge areas of coastal hills after fire. Also part of this family is the yellow monkeyflower *(Mimulus brevipes),* an erect annual 12 to 30 inches in height that can commonly be found on dry chaparral slopes after fire or other disturbance (pl. 145).

Although not specifically a fire follower, one of the most spectacular herbs that may be encountered on dry chaparral slopes is scarlet larkspur *(Delphinium cardinale).* In the improved growing conditions after fire this perennial species often reaches six to eight feet in height with stems covered in bright scarlet flowers (pl. 146). Early travelers to California described masses of flowering scarlet larkspur as giving the appearance of a hillside on fire.

Less Showy
Fire-Following Annuals

Not all fire-following annuals are large and showy. Among the less prominent are two species of the borage family (Boraginaceae). This is also called the forget-me-not family after one of its best-known members. Like the phacelias described above, the borages have radially symmetrical flowers arrayed along coiled stems. The flowers, however, are commonly small and much less showy than those of the phacelias. The stems have dense stiff hairs. Borages are a large and diverse group of annuals in California but are more typical of grasslands or open dry habitats. The two widespread postfire borages have multiple common names but are widely known as the large-flowered cryptantha *(Cryptantha intermedia)* and the small-flowered cryptantha *(C. microstachys)*. As their common names suggest, they are most easily separated on the basis of their flower size. The large-flowered cryptantha has flowers one-eighth to one-fourth inch in diameter (pl. 147),

Plate 147. Large-flowered cryptantha (*Cryptantha intermedia*, Boraginaceae), March to July.

Plate 148. Small-flowered cryptantha (*Cryptantha microstachys*, Boraginaceae), April to July.

whereas the small-flowered cryptantha has tiny flowers only about one-thirty-second inch in diameter (pl. 148). Despite the small flowers of both species, dense masses of these and other cryptanthas can collectively be quite showy. The early Spanish colonists in California called them *nievitas* because fields of these flowers resemble freshly fallen snow. The stems of a number of cryptantha species and the related popcorn flowers (*Plagiobothrys* spp.) contain a sap that can stain hands or pants a rich purple.

Annual species in the phlox family (Polemoniaceae) are prominent fire followers with small but beautiful flowers. Among these is the globe gilia *(Gilia capitata),* sometimes called blue-headed gilia. This is a tall, slender annual that may reach nearly three feet in height, although it is often much shorter. It has 10 to 50 or more pale blue violet flowers clustered in a dense spherical head at the end of a long naked stem (pl. 149). The leaves are doubly dissected into small linear lobes. Globe gilia is most abundant after fires but may be found more widely on dry chaparral slopes. Splendid gilia *(G.*

Plate 149. Globe gilia (*Gilia capitata* subsp. *abrotanifolia*, Polemoniaceae), April to May.

australis), not pictured here, is a small annual with a basal rosette of highly dissected leaves; it rarely grows as tall as 12 inches.

Most of our sage species in Southern California are small shrubs or herbaceous perennials, but one is a characteristic fire-following annual in chaparral. This is chia *(Salvia columbariae)*, a widespread species that was an important food plant for Native Americans. Chia grows to variable heights ranging from a few inches to two feet and bears clusters of bluish purple flowers in separate whorls along the flowering stems (pl. 150). Like the whispering bells described above, chia is an ecologically variable species. Chaparral populations germinate in huge numbers in the first spring following a fire, whereas desert populations germinate abundantly in the absence of fire in years with good rains. Although chia seeds are tiny, they are rich in nutritious oils and mucilage. Native Americans gathered chia seeds in huge quantities by bending the mature

Plate 150. Chia (*Salvia columbariae*, Lamiaceae), April to May.

heads over flat, tightly woven baskets and beating them until the seeds dropped. The seeds were consumed in a variety of ways. Most commonly, they were dried and ground in rock mortars into a fine flour. This could be eaten directly or mixed with water to form a thick gruel. Legend says that a single teaspoon of seeds could keep a warrior satisfied for a day of forced march. The unground seeds could also be mixed with water, with the mucilage dissolving to form a refreshing drink. The early Spanish settlers in California also used chia to make a nutritious beverage, adding lemon and sugar for flavoring.

Postfire Shrubs and Subshrubs

In the classic story of succession or recovery after fire in chaparral stands, the previously dominant shrub species return to controlling the canopy cover within a few years as the root crowns resprout and new seedlings germinate. In Southern California, however, there is an intermediate stage: a number of short-lived shrubs and semiwoody species, or subshrubs, become established in large numbers after fire from seed stored in the soil. These remain relatively dominant for two to

Plate 151. Deerweed (*Lotus scoparius,* Fabaceae), all year.

six years or more, until the original woody shrub species reestablish their canopy cover.

The most widespread of these postfire species, and certainly the most significant from an ecological standpoint, is deerweed *(Lotus scoparius)*. Deerweed is a semiwoody member of the legume family (Fabaceae). It grows to three to four feet or more in height but is only truly woody at the base of its main stem. Although it possesses scattered small trifoliate leaves, its green stems furnish much of its photosynthetic surface (pl. 151). These stems maintain active photosynthesis far into the dry summer months, long after the leaves have been shed. The small flowers, which resemble those of the pea, are yellow but turn reddish with age. They are borne in groups of one to four on very short stalks whorled from the leaf axils. Surprisingly, deerweed can be found flowering in any month of the year if sufficient moisture is present.

Like many legumes, deerweed possesses large seeds with a thick seed coat. These remain dormant over long periods until something happens to crack or abrade the coat so that water can enter the embryo. For exposed seeds, this event can be weathering, but for buried seeds it is commonly the heat of a chaparral fire. Large numbers of these seeds germinate after fire, and the plants grow quickly. Although the total mass of all of the annuals typically exceeds that of deerweed in the first year after a fire, deerweed typically forms the greatest biomass in these stands from the second through the third or fourth year, when the resprouting shrubs begin to reestablish their dominance. This abundance of deerweed is both due to and significant because of the ability of this species to fix nitrogen using symbiotic bacteria in special root nodules. This is the same process used by shrubby legumes such as chaparral pea *(Pickeringia montana)*, as described previously. In much the same way that legume crops such as clover and alfalfa increase soil fertility, deerweed adds large amounts of nitrogen to chaparral soils that have lost this critical element in gases released by the heat of the fire. Without deerweed to supply this nitrogen, frequent fires would steadily deplete the amount of nitrogen available for plants in chaparral soils.

Deerweed plays another ecological role as well: its foliage is high in protein and thus desirable forage for a variety of animal species, such as deer. Mature chaparral, with its tough leathery leaves, is a poor food source for browsing animals, but chaparral communities in the first few years following a fire provide a wealth of high-quality forage. The Native Americans recognized this and used fire widely to increase favored browse for deer and other game animals. Deerweed is also an important nectar plant for a variety of insect species.

Another common postfire shrub, especially on north-facing slopes, is the bush poppy *(Dendromecon rigida),* our only woody member of the poppy family. The large yellow flowers of bush poppy, up to three inches across, have the four petals (rarely five) and numerous stamens typical of members of the poppy family (pl. 152). Unlike those of its herbaceous relatives, the elongate, pale gray green leaves of bush poppy are evergreen and leathery in texture. The seeds of bush poppy are dispersed by ants; such dispersal is rare in the

Plate 152. Bush poppy (*Dendromecon rigida,* Papaveraceae), February to April.

California flora. Harvester and carpenter ants are attracted by an oil-rich body attached to the seed coat. The seeds germinate in large numbers after chaparral fires, and the plants quickly grow to heights of six to 10 feet. Such massive establishment was apparent in many parts of the Santa Monica Mountains following the large fires of 1993. Although these shrubs appear solid and woody, their stems are filled with soft tissue and can be easily broken. Bush poppy is not a long-lived shrub. By five to six years after the 1993 fires, the original chaparral shrub cover had returned to dominance and the bush poppy had begun a period of slow decline.

Another shrub species that often becomes established in dense thickets after chaparral fires is bush mallow (*Malacothamnus fasciculatus*). Rather than being strictly tied to fire, however, bush mallow is a colonizer after any form of disturbance (pl. 153). Although relatively short lived compared to typical chaparral shrubs, bush mallow, with its wandlike branches, can reach heights of up to 15 feet and remain persistent in chaparral openings for many years.

Short-lived but perennial relatives of bush poppy are golden ear-drops *(Dicentra chrysantha)* and cream-colored ear-drops *(D. ochroleuca),* which have other evocative common names such as bleeding hearts and fire hearts. These names all come from the flattened heart-shaped flowers, which are formed of four petals, an outer larger pair and an

Plate 153. Bush mallow (*Malacothamnus fasciculatus*, Malvaceae), April to October.

inner spoon-shaped pair that form a dome over the stamens. Both species of ear-drops are common on north-facing slopes after fires or other disturbances. A basal rosette of highly divided leaves forms the first year, and a tall flowering stalk reaching to six feet is topped by dense clusters of flowers. The northern half of our region has cream-colored ear-drops which, as the name suggests, has cream-colored flowers (pl. 154). Golden ear-drops, which has bright yellow flowers, is a more widespread species and dominates in San Diego County.

One final species of herbaceous perennial that often establishes from seed after fires is wild morning-glory *(Calystegia macrostegia)*. This scrambling perennial vine with stems trailing for up to 10 to 12 feet is found irregularly in disturbed sites on dry chaparral slopes, but often germinates in incredible numbers on rocky south-facing slopes after fires or other

Plate 154. Cream-colored ear-drops (*Dicentra ochroleuca*, Papaveraceae), May to July.

Plate 155. Wild morning-glory (*Calystegia macrostegia* subsp. *cyclostegia*, Convolvulaceae), February to May.

disturbances in chaparral or coastal sage scrub. The large white flowers have five petals joined together to form a trumpet-shaped tube, with purple stripes on the outer surface (pl. 155). The distinctive arrowhead-shaped leaves are one to two inches in length.

Woodlands, particularly those dominated by oaks, form a characteristic element of the vegetation of Southern California. They include closed-canopy woodlands of evergreen oaks, open savannas of deciduous oaks in a matrix of grassland cover, walnut woodlands, and local areas of conifer woodlands. Very often, woodlands and chaparral communities occur in a mosaic. Chaparral communities are generally associated with shallow or rocky soils, woodlands with deep soils or fractured geological substrates that allow roots to penetrate deeply and tap groundwater pools. Five forms of woodlands are considered here.

Live Oak Woodlands

Woodlands dominated by coast live oak *(Quercus agrifolia)* form a characteristic community on north-facing slopes of the foothills throughout our region and in shaded canyons where groundwater is available at depth (pl. 156). Unlike the open woodlands and savannas with grassy understories that are typical of the Engelmann oak and valley oak communities described below, live oak woodlands often have closed canopies of oak and tall shrubs. The characteristic herb cover in the shade of these stands is largely lacking in grasses and very different from that of the open savannas and grasslands.

Coast live oaks are large trees, with canopies that reach to heights of 60 feet and trunk diameters that can span eight to 10 feet (pl. 157). These massive trees often have large gnarled branches diverging both vertically and horizontally to produce a rounded, unusually broad canopy that can extend well over 100 feet across. Their dense evergreen canopy of leaves and their twisted complex of massive branches led many early settlers to write about the mystic beauty of these trees. It is easy to capture this feeling when you walk through a stand of coast live oaks in the early morning or just before sunset,

Plate 156. Coast live oak woodland.

when the low angles of the sun cast a myriad of shadows and present a haunting beauty.

With their deep root systems, coast live oaks thrive anywhere that they can penetrate deeply into the soil and rocky substrate to tap groundwater pools. Though they favor north-facing slopes and shaded canyons, coast live oaks also contribute much of the tree canopy cover along stream margins in Southern California. They are not true riparian trees, however. These oaks never root in the stream channel, as they cannot tolerate flooded soils, and they tap underground water supplies rather than stream water. Although they cannot tolerate salt spray or

Plate 157. Coast live oak (*Quercus agrifolia,* Fagaceae), March to April.

ocean winds, coast live oaks are unusual among California oaks in their ability to thrive on coastal slopes near the ocean.

Coast live oak woodlands often merge with mixed chaparral on north-facing slopes in Southern California, as described in chapter 5. These stands typically have an upper canopy of live oak with an understory of large woody shrubs in the light gaps at the edges of the canopies. The most common species of shrub in these stands is hollyleaf redberry *(Rhamnus ilicifolia)*, which grows best at the edges of canopies of coast live oak. Other chaparral shrub species commonly found mixed with coast live oak are toyon *(Heteromeles arbutifolia)*, hollyleaf cherry *(Prunus ilicifolia)*, and greenbark ceanothus *(Ceanothus spinosus)*. Less common but not unusual in open areas of live oak woodland are chaparral shrubs such as laurel sumac *(Malosma laurina)*, sugar bush *(Rhus ovata)*, and California mountain mahogany *(Cercocarpus betuloides)*.

Two small tree species widely present in live oak woodlands are California walnut *(Juglans californica)* and Mexican elderberry *(Sambucus mexicana)*. California walnut can occur as a dominant tree in its own woodlands, as described below, and can often be found growing with coast live oak in shaded canyons (pl. 158). Mexican elderberry is a shrubby winter-deciduous tree that favors open sites within live oak woodlands (pl. 159).

The most common understory species in moist oak woodlands, much to the discomfort of many hikers, is poison oak *(Toxicodendron diversilobum)*. This species, famous for producing severe skin rashes, is remarkably plastic in its growth form. It is found most commonly as a sprawling semiwoody shrub in the shaded understories of these woodlands, but at times it takes on the form of a woody vine extending up to 40 feet or more into oak canopies. Poison oak is easy to recognize in spring and summer; its three leaflets have an oily surface appearance (pl. 160). These leaves take on a reddish hue in fall and are lost over the winter. In its leafless condition poison

Plate 158. California walnut (*Juglans californica*, Juglandaceae), April to May.

Plate 159. Mexican elderberry (*Sambucus mexicana*, Caprifoliaceae), April to August.

oak is difficult to recognize, but skin contact with the light brown stems with a cover of short hairs can cause the same bad reaction as the leaves.

Plate 160. Poison oak (*Toxicodendron diversilobum,* Anacardiaceae), March to April.

Live oak woodlands contain a number of less annoying species of understory shrubs. Perhaps the showiest of these and the largest, although it dies back strongly in summer, is canyon sunflower *(Venegasia carpesioides).* This perennial shrub with large yellow sunflowers reaches as much as six to eight feet in height but is truly woody only at its base (pl. 161). Canyon sunflower occurs abundantly in the understories of coast live oaks on north-facing slopes, in shaded canyons, and on stream banks. It is widespread in the Santa Monica Mountains and the northern part of our region and increases greatly in abundance after fires, but is much less common in San Diego County.

Snowberry *(Symphoricarpos mollis)* is a low spreading shrub one to two feet in height that occurs commonly in semishaded oak understories. This species is named for its white fruits, which remain on the plant well into winter, long after the leaves have been lost. Small oval leaves arranged oppositely along the stems and clusters of small pink bell-shaped flowers in spring readily identify snowberry (pl. 162).

Plate 161. Canyon sunflower (*Venegasia carpesioides,* Asteraceae), March to June.

Plate 162. Snowberry (*Symphoricarpos mollis,* Caprifoliaceae), April to May.

Golden currant *(Ribes aureum),* a yellow-flowered species of deciduous shrub, is another common understory species (pl. 163). It is our only *Ribes* with yellow flowers. Its close relative chaparral currant *(R. malvaceum)* (see pl. 110) is more typical of chaparral communities but may also be found in open areas of live oak woodlands. Oak gooseberry *(R. quercetorum)* is a widespread shrub with spiny stems (pl. 164).

White nightshade *(Solanum douglasii)* and purple nightshade *(S. xantii)* are semiwoody shrubs frequently encountered in the understories of live oak woodlands. These are relatives of our domestic tomatoes and potatoes, as seen in their flowers, which have a prominent cone of five yellow anthers closely surrounding the female stigma. White nightshade is

Plate 163. Golden currant (*Ribes aurem* var. *gracillimum*, Grossulariaceae), January to April.

Plate 164. Oak gooseberry (*Ribes quercetorum*, Grossulariaceae),
March to May.

the taller of the two, reaching three to six feet in height, and
has white flowers with greenish spots at their base (pl. 165). It
is largely restricted to shady woodland understories. Purple
nightshade is two to three feet in height and can grow in more

Plate 165. White nightshade (*Solanum douglasii,* Solanaceae), February to March.

open habitats. As its name suggests, it has showy purple flowers (pl. 166). Like many nightshades, these species are highly poisonous.

There are two additional common colonists of open areas or disturbed sites within live oak woodlands, as well as moister sites in chaparral. The more abundant of these and often the first species to flower anywhere in early winter is wild cucumber *(Marah macrocarpa),* a trailing herbaceous vine (pl. 167). Wild cucumber has a large tuberous root that can weigh 100 pounds or more, giving

Plate 166. Purple nightshade (*Solanum xantii,* Solanaceae), January to May.

Plate 167. Wild cucumber (*Marah macrocarpa*, Cucurbitaceae), January to April.

rise to another common name, manroot. As early as December each year, new stems sprout and grow quickly from carbohydrates stored within this tuber, clambering as much as 20 feet or more over shrubs and into low tree branches. The common name wild cucumber comes from the four-inch spiny fruits, which resemble cucumbers.

With the exception of poison oak, woody vines climbing into tree canopies are rare in Southern California. Two other species of woody vines, however, occur widely in coast live oak woodlands. Chaparral honeysuckle *(Lonicera subspicata)* is an evergreen vine that is common on the margins of live oak woodlands and taller stands of mixed chaparral (pl. 168). It can climb 30 feet or more into oak canopies but generally remains in a somewhat shaded position. The flowers, which bloom in late spring, are notably fragrant. Also present as a woody vine in oak woodlands, but largely invisible much of the year, is virgin's bower *(Clematis ligusticifolia),* which may

Plate 168. Chaparral honeysuckle (*Lonicera subspicata*, Caprifoliaceae), April to May.

climb 30 feet or more into oaks. Virgin's bower is characterized by oppositely arranged leaves, each dissected into five to seven or more leaflets (pl. 169). This species is most apparent not when it is flowering, but rather when long plumes of the mature seeds have twisted themselves in silver-colored silky balls. A closely related species that more characteristically climbs into chaparral shrubs in shaded canyons is chaparral clematis *(C. lasiantha)*, characterized by just three (rarely five) leaflets (pl. 170).

Several large perennial herbs are common and widespread in the understories of live oak woodlands in Southern California. Perhaps the most abundant of these in the northern part of our region is pitcher sage *(Salvia spathacea)*, a coarse low perennial with a square and woolly stem. Unlike the majority of its brethren, pitcher sage lacks a strong odor in its leaves. It is easily recognized when flowering by its two-lipped crimson flowers (pl. 171). These contain large amounts of

Plate 169. Virgin's bower (*Clematis ligusticifolia,* Ranunculaceae), April to June.

Plate 170. Chaparral clematis (*Clematis lasiantha,* Ranunculaceae), February to May.

nectar and are a favorite of hummingbirds. An unusual herbaceous perennial that is often among the first species to flower in midwinter is wild peony *(Paeonia californica).* This is a tall fleshy herb with a bushy growth form that reaches about two feet in height. It has large blackish red flowers that nod singly on smooth green stems and never fully open (pl. 172). A large fern species, coastal wood fern *(Dryopteris arguta),* also deserves mention in this category of large herbaceous perennials. This fern, common on moist slopes under

Plate 171. Pitcher sage (*Salvia spathacea*, Lamiaceae), March to May.

Plate 172. Wild peony (*Paeonia californica*, Paeoniaceae), January to March.

oak canopies, has a rosette of long fronds extending up to two feet or more (pl. 173). It is our only large fern found in such habitats.

Two large perennial grasses are common in open areas on moist slopes in oak woodlands, although they may also occur on chaparral slopes in moist local areas. The first and more apparent of these is giant rye grass *(Leymus condensatus)*, a tall perennial grass that resembles a small bamboo (pl. 174). This was an important plant to Native Americans in Southern California, who used the semiwoody stems to fashion arrow shafts. It is the only native grass to attain a height of six feet or more. It can only be confused with

Plate 173. Coastal wood fern (*Dryopteris arguta*, Dryopteridaceae).

Plate 174. Giant rye grass (*Leymus condensatus*, Poaceae), June to August.

giant reed *(Arundo donax)*, a far larger and woodier invasive grass common along stream courses. The other large perennial grass is deergrass *(Muhlenbergia rigens)*, a bushy bunchgrass reaching heights of four to five feet when blooming (pl. 175). In many respects, deergrass is a highly significant species for wildlife. Deer use clumps of deergrass for cover when they have young fawns, and many mammals graze on the young grass blades. The seeds provide food for many birds, and the plant itself is an important larval food source for several butterfly species. Native Americans used the flowering stems as the foundation for their famous coiled baskets, with an individual basket requiring thousands of these stems.

The understories of live oak woodlands display a wide diversity of showy annuals and small herbaceous perennials in spring. Some that occur in sunnier openings are also found widely in grasslands; many others are largely restricted to shaded sites beneath the oaks. Typically, the first of these to flower is milkmaids *(Cardamine californica),* often called toothwort (pl. 176). This perennial herb grows from under-

Plate 175. Deergrass (*Muhlenbergia rigens,* Poaceae), June to August.

Plate 176. Milk-maids (*Cardamine californica*, Brassi-caceae), February to April.

ground tubers and quickly sprouts and flowers as early as December if there have been fall rains.

A second wave of flowering begins in February or early March. Miner's lettuce *(Claytonia perfoliata)* is a common annual in this group. This is an easy species to recognize, with succulent stems and leaves that merge together to form a continuous whorl around the stem (pl. 177). Its common name comes from the fact that the young leaves are highly edible and were commonly eaten as a raw salad by both the Native Americans and the early European settlers. Also flowering in early spring are a group of annuals in the phacelia family (Hydrophyllaceae) that often grow side by side on moist slopes. Baby blue-eyes *(Nemophila menziesii)* is a distinctive member of this family. Its large saucer-shaped flowers have petals that are bright blue on the outer half and pale blue below (pl. 178).

Plate 177. Miner's lettuce (*Claytonia perfoliata*, Portulaceae), February to May.

Plate 178. Baby blue-eyes (*Nemophila menziesii,* Hydrophyllaceae), February to June.

Eucrypta *(Eucrypta chrysanthemifolia)* is a small annual with tiny white bell-shaped flowers in terminal clusters and small fernlike leaves (pl. 179), and fiesta flower *(Pholistoma auritum)* is a straggling herb that has purple flowers and coarse square stems with downward-pointing prickles along their margins (pl. 180). The common name of this flower apparently dates from the early days of the Spanish ranchos, when

Plate 179. Eucrypta (*Eucrypta chrysanthemifolia*, Hydrophyllaceae), February to May.

Plate 180. Fiesta flower (*Pholistoma auritum*, Hydrophyllaceae), March to May.

young women decorated their party dresses with stems of fiesta flower. The prickles allow the stems to readily stick to any fabric.

Several annual species growing within oak woodlands are notable for blooming later in the season. One of these is Chinese houses (*Collinsia heterophylla*), an upright annual with two-lipped flowers; the upper petals are white and the lower petals purple (pl. 181). The arrangement of the ascending whorls of these flowers makes the inflorescence resemble a

Plate 181. Chinese houses (*Collinsia heterophylla*, Scrophulariaceae), April to May.

Plate 182. Indian pink (*Silene laciniata* subsp. *major*, Caryophyllaceae), April to July.

Chinese pagoda. Also late in flowering is Indian pink *(Silene laciniata)*, a sticky and hairy small perennial with scarlet red flowers (pl. 182). Each of the petals of Indian pink is deeply cleft into four linear lobes.

Engelmann Oak Woodlands

Engelmann oak woodlands, named for the dominant presence of Engelmann oak *(Quercus engelmannii),* are one of the most threatened communities of oaks in California. These are open woodlands and savannas with grassy understories that historically ranged from the foothills of the San Gabriel Mountains near Pasadena through the Chino Hills and southward along the slopes of the Peninsular Ranges to San Diego County and northern Baja California. Encroaching urban development in recent decades, however, has eliminated this community from most of its northern and western range, and today 90 percent of existing stands occur in San Diego County. The only two large protected areas of Engelmann oak woodlands are on the Santa Rosa Plateau of western Riverside County (pl. 183) and the area around Black Mountain in the Peninsular Ranges of central San Diego County.

Engelmann oaks are relatively large oaks that reach to 40 feet in height, with gracious globe-shaped canopies, and have

Plate 183. Engelmann oak woodland.

Plate 184. Engelmann oak (*Quercus engelmannii,* Fagaceae), March to April.

trunk diameters of one to four feet. Their leathery blue green leaves are one to three inches long and lack the marginal teeth or lobes present in other tree oaks of our region (pl. 184). The leaves are generally described as semievergreen because some cohorts of leaves formed in spring typically remain on the tree until new leaves are formed the following year. Thus, Engelmann oaks are unusual among our oaks; all the rest either are fully winter deciduous or are evergreen and retain each cohort of leaves for two years.

Engelmann oaks may occur as the only dominant tree in their woodlands but are more commonly mixed in open stands with coast live oaks. They may also occur with willows (*Salix* spp.), sycamores (*Platanus* spp.), and cottonwoods (*Populus* spp.) adjacent to stream corridors, but not within the riparian zone itself. In some areas of San Diego County, Engelmann oak stands may occur adjacent to communities of chamise chaparral. Large Engelmann oaks are thought to range from 100 to 150 years in age, although there are suggestions that a few trees may reach ages of three centuries or more.

Deciduous Oak Woodlands and Savannas

Valley oak savannas are most typical of the Coast Ranges of central California and extend into the valleys and inner drainages of the Coast Ranges of Santa Barbara County and from the Tehachapi Mountains southward into the Santa Clarita Valley of Los Angeles County (pl. 185). Valley oaks *(Quercus lobata)* reach their southern limit on the northern slopes of the Santa Monica Mountains from Thousand Oaks to Calabasas. Valley oaks are often considered to be the monarchs of California oaks because they can grow larger than any other oak species. Large trees have massive trunks, often six to eight feet in diameter, and are as much as 100 feet tall. Valley oaks are winter-deciduous trees with round spreading canopies and pendulous smaller branches that droop or weep downward. The large leaves are two to four inches in length and deeply lobed into rounded segments lacking sharp points (pl. 186).

Plate 185. Valley oak woodland.

Plate 186. Valley oak (*Quercus lobata,* Fagaceae), March to April.

An important limiting factor for these trees is the need for deep soil profiles that the extensive root systems can explore for moisture. Thus, they prefer bottomlands and areas with groundwater supplies of moisture. Low rainfall, shallow soils, or both keep valley oaks from moving farther south in our region. All of their southern populations have been strongly impacted by urban expansion into the broad valleys where they occur. The rapid expansion of vineyards in Santa Barbara County in recent years has also had major impacts on this species.

A serious problem in valley oak savannas throughout California, including our area at the southern end of the range of the species, is a general lack of successful establishment of sufficient seedlings to maintain adult populations. Although mature valley oaks are thought to reach ages up to 300 years or more, few saplings and seedlings are present in most stands. This lack of successful reproduction is thought to be due to a variety of causes, including seedling competition with invasive grasses, predation by diverse vertebrates, and genetic problems caused by reduced outcrossing of isolated trees.

Plate 187. Blue oak woodland (photograph by Philip W. Rundel).

Blue oak woodlands are widespread in the Coast Ranges of central and northern California and the Sierra Nevada foothills (pl. 187) but only reach the northern limits of our region. These deciduous oak woodlands are important in the southern Coast Ranges of Santa Barbara County and in the Tehachapi Mountains in the northwestern corner of Los Angeles County. Their dominant species is blue oak *(Q. douglasii)*, a winter-deciduous species with shallowly lobed leaves that are distinctly blue green in color (pl. 188). Often associated with blue oak are California buckeye *(Aesculus californica)* (pl. 189) and interior live oak *(Q. wislizenii)* (pl. 190). This latter shrubby tree extends southward into the Transverse and Peninsular Ranges.

Nonnative annual grasses, as discussed in the next chapter, form the understory matrix of cover in open stands of both valley and blue oak. Introduced from the Mediterranean Basin by the early Spanish settlers beginning two centuries

Plate 188. Blue oak (*Quercus douglasii*, Fagaceae), March to April.

Plate 189. California buckeye (*Aesculus californica*, Hippocastanaceae), May to July.

Plate 190. Interior live oak (*Quercus wislizenii*, Fagaceae), March to May.

ago, these grasses have now almost entirely replaced our native perennial bunchgrasses as the dominant herb species in these communities. Native and alien grassland species common in these open woodlands and savannas are described in chapter 7.

Walnut Woodlands

Although typically an associate of coast live oak in shaded canyons, California walnut is dominant in some woodlands (pl. 191). The most extensive of these stands can be found on north-facing slopes of inland valleys in Ventura, Los Angeles, and northern Orange Counties. The stands with the highest cover occur in the Santa Clara River basin; on the north slopes of the Santa Monica Mountains; and in the Simi, San Jose, Puente, and Chino Hills. Farther south these trees occur infrequently. Although often relatively small, California walnut can reach 30 to 40 feet in height and two to four feet in diameter under favorable conditions. The only common associated tree species in walnut woodlands are coast live oak and Mexican elderberry. The soft wood of California walnuts readily forms cavities that are important nesting sites for many bird species. The hard-shelled walnuts are an important food resource for both ground squirrels and Western Gray Squirrels *(Sciurus griseus)*. For humans, a considerable

Plate 191. California walnut, a small woodland tree that can grow to large size along streams.

amount of effort is necessary to extract the edible meat from the shells. Handling the walnuts can produce a black stain that is difficult to remove from clothing.

Conifer Woodlands

Although forests of pines and firs form the dominant vegetation above about 6,000 feet in the mountains of Southern California, conifers are relatively rare and scattered at lower elevations. At the end of the last Ice Age, pine woodlands covered extensive areas along our coasts, and coast redwood *(Sequoia sempervirens)* was present in the moist canyons of the Santa Monica Mountains. However, the climate warmed and became more arid, so these forests were largely lost. The southern limit of coast redwoods is now in San Luis Obispo County, and coastal pine forests today occupy only scattered remnants of their former range.

Plate 192. Torrey pine (*Pinus torreyana,* Pinaceae).

The most famous of the coastal pine relicts in Southern California is Torrey pine *(Pinus torreyana)*, known from just two isolated stands (pl. 192). One of these stands, largely protected in Torrey Pines State Park, lies on the coast near Del Mar to the north of San Diego. Frequent coastal fogs are associated with this site and possibly provide extra moisture for these pines through fog drip onto the soil. Surprisingly, how-

Plate 193. Bishop pine (*Pinus muricata*, Pinaceae).

ever, this coastal site is also home to a number of shrub and succulent species more typical of desert habitats. The second population of Torrey pine lies on Santa Rosa Island. This five-needled pine tolerates dry, nutrient-poor soils and responds well to fire. Poor regeneration in the coastal population may be related to fire protection.

Our second coastal pine, also rare in our area but common along the coast of northern California, is Bishop pine *(P. muricata)*. This two-needled species is common on the coast of northern California but is rare in our area, present only as scattered individuals in the coastal hills of Santa Barbara County and on Santa Rosa and Santa Cruz Islands (pl. 193). There is a southern relict population near San Vicente in northwestern Baja California, 250 miles south of any other stand.

Two foothill pine species can also be mentioned briefly here. Coulter pine *(P. coulteri)* is found in chaparral and open

Plate 194. Coulter pine (*Pinus coulteri*, Pinaceae).

oak woodlands in many foothill areas of the higher mountain ranges of Southern California, extending upward from about 2,500 feet into the lower forests of ponderosa pine *(Pinus ponderosa)*. Coulter pine is a medium-sized pine that can reach heights of 70 to 80 feet and diameters of two to three feet. It can be recognized by long, stiff needles in bundles of three and by very large globular cones (pl. 194). These cones can commonly reach 12 inches or more in diameter and weigh up to five pounds or more when green, making them the largest cones of any pine in the world. Although it is widely distributed across the foothills of the Transverse and Peninsular Ranges, Coulter pine is inexplicably absent from the Transverse Ranges of western Los Angeles and Ventura Counties despite the presence of seemingly appropriate habitat.

Gray pine *(P. sabiniana)*, sometimes called foothill or digger pine, is the most widespread foothill pine in California

Plate 195. Gray pine (*Pinus sabiniana,* Pinaceae).

but only barely enters our area in the Tehachapi Mountains of northern Los Angeles County. This is an easy species to recognize by its long, wispy gray green needles in bunches of three and most particularly by its frequent branching and common lack of a single main trunk (pl. 195).

Another foothill conifer of interest is bigcone Douglas-fir *(Pseudotsuga macrocarpa),* sometimes called bigcone spruce (pl. 196). Bigcone Douglas-fir has a scattered occurrence on rocky slopes above 3,000 feet in the Transverse and Peninsular Ranges, where it extends from chaparral communities up into the lower mixed conifer forests. It often occurs either as isolated trees in mixed chaparral or on rocky sites associated with canyon live oak *(Quercus chrysolepis).* Although bigcone Douglas-fir is able to survive only light fires, a seemingly poor trait for a tree growing within flammable chaparral, it thrives on rocky sites and in canyons, where intense fires do not penetrate. These protected sites allow this species to reach ages of up to 800 years. In general appearance, bigcone Douglas-fir resembles Douglas-fir *(Pseudotsuga menziesii),* a closely re-

Plate 196. Bigcone Douglas-fir (*Pseudotsuga macrocarpa*, Pinaceae).

lated tree widespread in montane habitats of the western United States. Most of Southern California is too dry for Douglas-fir, although its southernmost point of occurrence in California is just within our region, in the Purisima Hills of Santa Barbara County.

Two rare species of cypress form low woodlands within stands of mixed chaparral on Cuyamaca Peak in San Diego County. Tecate cypress *(Cupressus forbesii),* a rare shrubby conifer 20 to 30 feet in height (see pl. 128), occurs from the Santa Ana Mountains southward into northwestern Baja California as far as San Vicente. The second species, Cuyamaca cypress *(C. arizonica),* is found only on Cuyamaca Peak in Southern California but occurs more broadly outside of our area. These cypress species generally require fire to open their sealed cones and release seed, although intense fires kill the parent trees. Abundant seedlings replace the parents in the population, but a second fire before the seedlings mature may destroy the population entirely.

Historical Changes

Grasslands in California once likely consisted of a mixture of
native perennial bunchgrasses, annual grasses, and broad-
leaved herbs—but we don't really know. The advent of the
first European settlers in the eighteenth century began the
rapid introduction and establishment of nonnative annual
grasses from the Mediterranean Basin. Some of these grasses
were deliberate introductions; others were probably hitch-
hikers in animal feed or in the fur of sheep or cattle brought
from Spain. Whatever their origin, these grasses and some
ecologically similar broad-leaved herbs from the Mediterranean Basin rapidly expanded within decades to take over
dominance of grasslands in California from the native species
(pl. 197). We lack precise records of when some of these annual grasses were first introduced because there were no
botanists here at that time. Straw used to make adobe bricks

Plate 197. Grassland community on the Santa Rosa Plateau with native
bunchgrasses retaining significant cover.

for some of the earliest California missions, however, contains fragments of many species of these invasive grasses.

Grasslands in Southern California today are largely dominated by a group of nonnative Mediterranean annual grasses. The most important of these are wild oats *(Avena barbata* and *A. fatua)* (pl. 198, *top left*), red brome *(Bromus madritensis)* (pl. 198, *top center*), ripgut grass *(B. diandrus)* (pl. 198, *top right*), soft cheat *(B. hordeaceus)* (pl. 198, *middle left*), barley *(Hordeum murinum)* (pl. 198, *middle right*), and Italian ryegrass *(Lolium multiflorum)* (pl. 198, *bottom right*), although many other alien species may be present.

There is no simple explanation as to why annual grasses from the Mediterranean Basin were able to replace our native grassland species so rapidly. Clearly, the European grasses arrived well adapted to spread and colonize disturbed areas quickly as a result of thousands of years of evolution in association with the earliest agricultural civilizations in the eastern Mediterranean Basin. Our best guess is that a combination of environmental disturbances enabled these changes to take place. Heavy grazing pressure from domestic animals began with the Spanish missions and accelerated rapidly in the gold rush era as large ranchos in Southern California found a ready market for beef cattle. The arrival of European settlers also brought a change in the natural fire regime of California grasslands. A reduction in range fires under human management of rangelands would have led to a buildup of dense layers of mulch, a condition that favors annual grasses over the native bunchgrasses. During the nineteenth century California also suffered several periods of extreme drought, a condition that annual grasses are much better able to survive than native bunchgrasses. Since becoming well established, the invasive annual grasses have never relinquished their dominance to the native grassland species.

The historical conversion of grasslands from perennial

bunchgrasses to annual grasses has had profound effects on the ecology of these communities. One effect has been to dramatically alter the seasonal pattern of water availability in the upper soil horizons. Shallow-rooted annual grasses grow luxuriantly through the spring months and use their extensive mats of fibrous roots to extract all of the available moisture from the upper soil horizons before the onset of their senescence. Mature perennial bunchgrasses are much more deeply rooted and thus drought resistant. With strong competition from annual grasses for available moisture in upper soil horizons, however, bunchgrass seedlings have difficulty surviving their first summer drought. Annual grasslands reduce the survival of seedlings of oaks and other woody plants in the same manner.

It is unclear whether large areas of native grassland without scattered tree cover were present before the advent of European settlers. Many of the areas in our region that are dominated by annual European grasses today almost certainly once supported communities of woody vegetation. Nonnative grasslands are extensive today in the Simi Hills of Los Angeles County, portions of central Orange County, the Perris Plain of Riverside County, and Camp Pendleton in San Diego County. It seems quite likely that coastal sage scrub or chaparral communities originally covered many of these areas. Their conversion to grasslands has taken place as a result of heavy grazing and changes in fire frequency over the past two centuries. We know that fires at intervals of one or two years can eliminate shrub cover in chaparral or coastal sage scrub and convert these areas to alien grasslands by preventing the successful establishment of resprouts and seedlings. This process can be readily seen along Hwy. 14 between Santa Clarita and Palmdale and along Hwy. 60 east of Riverside, where communities of interior sage scrub have largely been converted into annual grasslands. One sage scrub species that survives well under these conditions, however, is the chaparral yucca (*Yucca whipplei*) (see pl. 53).

Wild oats (*Avena fatua,* Poaceae), April to June.

Red brome (*Bromus madritensis* subsp. *rubens,* Poaceae), April to June.

Ripgut grass (*Bromus diandrus,* Poaceae), April to June.

Soft cheat (*Bromus hordeaceus,* Poaceae), April to July.

Barley (*Hordeum murinum* subsp. *leporinum,* Poaceae), April to June.

Plate 198. Nonnative European grasses.

Italian rye-grass (*Lolium multiflorum,* Poaceae), June to August.

Native Grasslands
on Nutrient-Poor Soils

Local areas of the Santa Monica Mountains and the Coast and Peninsular Ranges retain stands of what appear to be natural perennial grasslands. Although nonnative annual grasses and other herbs are present, native bunchgrasses form the dominant cover. It appears that unusual soil conditions are often responsible for the success of native bunchgrasses. Heavy clay soils, in particular, often require special forms of adaptation for survival. These soils shrink and expand significantly with the seasons, making it difficult for the root systems of woody plant species to colonize these sites. Moreover, many invasive annual grasses require moderate to high levels of soil fertility and thus grow and reproduce poorly on nutrient-deficient soils. Native annuals and perennial geophytes (species that grow annually from underground bulbs, corms, or rhizomes), including a number of local endemic species, are often relatively abundant on these clay or other unusual soils.

Grassland Floras

A grassland understory of some form has likely always been associated with Engelmann oak and valley oak savannas in Southern California. The original dominants of this community, before the arrival of the European invasives, were native bunchgrasses, particularly giant stipa *(Achnatherum coronata)*, nodding needlegrass *(Nassella cernua)* (pl. 199), and purple needlegrass *(N. pulchra)* (pl. 200). These bunchgrasses remain today but are much reduced in abundance. They can be best seen on nutrient-poor or heavily eroded soils where the invasive annual grasses grow less aggressively.

Beyond the perennial bunchgrasses, diverse assemblages of native annuals and herbaceous perennials can be found

Plate 199. Nodding needlegrass (*Nassella cernua*, Poaceae), April to May.

Plate 200. Purple needlegrass (*Nassella pulchra*, Poaceae), March to May.

scattered among the invasive grasses and on occasion can cover huge areas where thinner soils restrict the invaders. Coast goldfields *(Lasthenia californica)* is a slender annual whose abundance in spring can spread a low carpet of gold over hillsides and fields, as its name suggests (pl. 201). The jack of spades in decks of cards used by early Spanish settlers in California is reported to have had one of these flowers in his hand. Young women referred to this species as *si me quieres, no me quieres* (love me, love me not), using its petals in this game as petals of daisies have been used for centuries. The same grasslands that hold expanses of goldfields often have two other species of annuals that provide additional splashes of color. California poppy *(Eschscholzia californica)*, our state flower, blooms abundantly in spring in many grasslands. It regularly forms massive displays covering hundreds of acres in disturbed grasslands at the edge of the desert in the Antelope Valley, the home of the famous California Poppy Reserve. Because of the tendency of the poppy flowers to close at night, the early Spanish settlers called this species *dormidera* (the drowsy one).

Plate 201. Coast goldfields (*Lasthenia californica,* Asteraceae), March to April.

Plate 202. Succulent lupine (*Lupinus succulentus*, Fabaceae), February to April.

Many annual lupine species can be found in grassy areas in Southern California. One of the most common is succulent lupine *(Lupinus succulentus),* which is six to 24 inches tall with blue or purple flowers and relatively coarse fleshy stems lacking hairs (pl. 202). Often growing in the same fields is dove lupine *(L. bicolor),* a small, delicate species rarely reaching more than 12 inches in height. The flowers of dove lupine are distinctly blue and white; the stems are hairless, but the leaves are covered with soft hairs (pl. 203). Both succulent lupine and dove lupine also appear frequently on burned chaparral slopes after fires. Collar lupine *(L. truncatus)* is another common annual grassland species. It gets its name from its linear leaflets, which are cut off squarely, or truncated, at their tips (pl. 204).

Plate 203. Dove lupine (*Lupinus bicolor,* Fabaceae), March to May.

The northern parts of our region often have large expanses of sky lupine *(L. nanus)* mixed in a patchwork mosaic with coast goldfields and California poppies. These lupines are medium-sized annuals six to 24 inches in height. They have the palmate divided leaves characteristic of all lupines, hairy stems, and whorls of bluish flowers (pl. 205). Coast goldfields, California poppy, and various lupine species together form remarkable splashes of spring color each year in the Tejon Pass area of the Tehachapi Mountains.

Spring fields of wildflowers in Southern California often have expanses of pinkish or reddish color scattered among the yellows, oranges, and blues of the goldfields, poppies, and lupines. These pinks and reds may indicate the presence of several different species. One of them, the ground pink *(Linanthus dianthiflorus),* can germinate in large numbers in wet years. In spring, this small annual, whose petals have

Plate 204. Collar lupine (*Lupinus truncatus*, Fabaceae), March to May.

Plate 205. Sky lupine (*Lupinus nanus*, Fabaceae), April to May.

Plate 206. Ground pink (*Linanthus dianthiflorus*, Polemoniaceae), February to April.

fringed outer margins (pl. 206), produces prodigious numbers of flowers on open slopes at low elevations.

Masses of mixed pinkish red and white flowers identify clusters of owl's clover *(Castilleja exserta,* formerly known as *Orthocarpus purpurascens),* one of a number of related species in our area. Up close these plants have dense clusters of small two-lipped flowers at the tops of their stems, with the upper lip crimson or purplish and the lower lip creamy white at its center, darkening to magenta on its margins (pl. 207). The leaves are highly dissected into many threadlike lobes. The name owl's clover comes from the fact that some observers see little owls in the shape of the white lower flower lips. The Spaniards called these plants *escobitas* or little whiskbrooms. This beautiful plant has a dark secret. Although it has green leaves and can photosynthesize, much of its nourishment comes from tapping its roots into those of

Plate 207. Owl's clover (*Castilleja exserta*, Scrophulariaceae), March to June.

Plate 208. Bird's beak (*Cordylanthus rigidus* subsp. *setiperus*, Scrophulariaceae), June to September.

surrounding species and parasitizing them for nutrients and sugars. A number of related grassland species are likewise what are called hemiparasites because of their mixed modes of nutrition. These include bird's beak *(Cordylanthus rigidus)*, a tall hairy annual (pl. 208), and Indian paintbrush *(Castilleja affinis)*, a semiwoody perennial.

Huge masses of white flowers covering grasslands in our region often mean that tidy tips *(Layia platyglossa)* are present. Tidy tips are mass-flowering annuals in the sunflower family (Asteraceae); the large flower heads have yellow ray flowers tipped with white (pl. 209). These distinctive flower heads are particularly attractive as well as fragrant. Tidy tips grow notably well on sandy flats along the coast, where they can color large areas, but they are very widely distributed in other areas. Millions of these flowers once covered extensive

Plate 209. Tidy tips (*Layia platyglossa,* Asteraceae), March to May.

expanses of the San Fernando Valley, but these populations have succumbed to urban expansion.

Many other species of native grassland herbs lack the mass flowering of the above species but nevertheless produce impressive blooms on a smaller scale. One of the first of these to bloom in spring is California buttercup *(Ranunculus californicus).* This slender perennial herb is six to 18 inches tall, with deeply lobed leaves and shiny yellow flowers possessing large numbers of both petals and stamens (pl. 210). Native Americans collected the seeds of this buttercup and ground them into flour for food. Cream-cups *(Platystemon californicus),* a member of the poppy family (Papaveraceae), are also early flowerers. These are small, delicate annuals with hairy leaves at the base of the plant and solitary cream-colored flowers at the ends of slender leafless stems (pl. 211). A third species flowering in early spring is common fiddleneck *(Amsinckia intermedia).* This is a tall, slender annual up to 30 inches in height, with bristly hairs on the stems and less densely on the linear leaves. The common name comes from the coiling of the small orange yellow flowers into an inflorescence shaped

Plate 210. California buttercup (*Ranunculus californicus,* Ranunculaceae), February to April.

like the neck of a violin (pl. 212). These are often abundant on dry grassy hillsides but can flower in greatly increased numbers after fires.

Not all of the showy grassland herbs are annuals. Blue larkspur *(Delphinium parryi),* a slender perennial herb with a fleshy root, reaches two to three feet in height in grassy areas throughout our region. It can be separated from a related species, spreading larkspur *(D. patens),* by its leaves, which are finely divided into linear lobes, and its relatively short flower stalks, which are no more than one-and-a-half inches in length (pl. 213). Spreading larkspur has broader leaf lobes, a thinner root, and flower pedicels one-and-a-half to four inches in length (pl. 214). It prefers semishaded grassland

Plate 211. Cream-cups (*Platystemon californicus*, Papaveraceae), April to May.

Plate 212. Common fiddleneck (*Amsinckia intermedia*, Boraginaceae), February to May.

Plate 213. Blue larkspur (*Delphinium parryi*, Ranunculaceae), June to July.

Plate 214. Spreading larkspur (*Delphinium patens* subsp. *hepaticoideum*, Ranunculaceae), March to May.

edges along woodland margins. Another showy herbaceous perennial is shooting star *(Dodecatheon clevelandii)*, a highly distinctive species. The unusual downwardly nodding pink flowers, looking much like small cyclamen flowers (pl. 215), are borne in a small group at the top of a naked stem several inches to a foot in height. Each flower has a yellow ring of color at its center and a bunched group of male stamens and female pistil extending outward like a purple beak. Shooting stars can be quite local in occurrence, often preferring areas of relatively wet soil.

A number of showy annual species mark the end of spring flowering. This late-flowering group is best exemplified by species of clarkias, sometimes called farewell-to-spring. The most common of these in our region is purple clarkia *(Clarkia purpurea)*. This species can reach 12 to 18 inches in height with slender stems displaying lavender to purple red flowers in the leaf axils (pl. 216). The flowers often have a dark

Plate 215. Shooting star (*Dodecatheon clevelandii*, Primulaceae), February to April.

purple spot on each petal. Very similar in form is farewell-to-spring *(C. bottae)*, which differs from purple clarkia in having flower buds that droop downward rather than standing erect (pl. 217). Two other common species of clarkias are quite distinct. The elegant clarkia *(C. unguiculata)*, a tall species that can reach heights up to three feet or

Plate 216. Purple clarkia (*Clarkia purpurea* subsp. *quadrivulnera*, Onagraceae), April to July.

more in years with good rainfall, has distinctive lavender pink to red purple petals that are broadly diamond shaped above but narrow to a slender base halfway down their length (pl. 218). Willow-herb clarkia *(C. epilobioides),* not pictured

Plate 217. Farewell-to-spring (*Clarkia bottae*, Onagraceae), April to June.

Plate 218. Elegant clarkia (*Clarkia un-guiculata*, Onagraceae), May to June.

here, is a slender annual with white flowers that turn pinkish with age. All of the clarkias share with the poppies the characteristic of having only four petals.

Two species of tarweeds, named for the sticky secretions produced along their stems, are common summer-flowering

Plate 219. Slender tarweed (*Hemizonia fasciculata,* Asteraceae), May to September.

Plate 220. Sticky madia (*Madia gracilis,* Asteraceae), May to June.

annuals in our grasslands. Slender tarweed *(Hemizonia fasci-culata)* is a tall, thin species reaching heights of up to three feet or more in open grasslands (pl. 219). The foliage has a disagreeable tarlike odor and the ability to stain clothing. The other common species of tarweed is sticky madia *(Madia gracilis),* a tall, slender annual with small heads of dense flower clusters (pl. 220).

Grassland Geophytes

Grasslands in Southern California are also home to a diverse variety of perennial geophytes. Many of these grassland geophytes were important food plants to the Native American tribes of our region, who gathered the bulbs or corms and roasted them. Indian women adept at using digging sticks could gather thousands of bulbs in a single afternoon.

Perhaps the most spectacular of these geophytes in our grasslands are species of mariposa lily, which can present large splashes of color on grassland slopes. There are more than 40 species of mariposa lilies in California, with about a third of these reaching our area. Most of these are characteristic of open grasslands or woodlands rather than chaparral, but many also can bloom in large numbers after chaparral fires. The common name mariposa comes from the Spanish word for butterfly. It describes the appearance of the large showy flowers, which seem to dance through the air on nearly invisible stems with even a slight breeze. All mariposa lilies have three showy petals, with the lower inside margin of each commonly showing a large colored spot marking the location of a nectary (a gland that secretes nectar). In addition to flower color and shape, the presence or absence of hairs on the inside surface of the petal and on the nectary itself are important characteristics in distinguishing species.

The Catalina mariposa lily *(Calochortus catalinae)* has large solitary flowers that are white with tinges of lilac (pl. 221, *top left*). The lower inside margin of each of the three petals displays a dark purple gland covered with coarse hairs. Although widespread in disturbed grasslands and on chaparral slopes after fire, this species prefers heavy clay soils. The lilac mariposa lily *(C. splendens)* is similar in general appearance but has deep lilac to pink or purple flowers (pl. 221, *top right*), whereas the butterfly mariposa lily *(C. venustus)* has white to purple flowers with a dark red blotch in the middle of each petal (pl. 221, *middle left*). The former is characteristic of clay soils, whereas the latter prefers sandy soils and reaches only the northern portions of our region. All three of these species lack hairs on the inside surface of the petals but have thick dense hairs on the large nectary at the inside base of each petal. Also similar is Weed's mariposa lily *(C. weedii),* whose flowers can vary in color from cream to purplish to deep yellow (pl. 221, *middle right*). It can be distinguished from other species by the lack of hairs on the nectaries and the presence of long yellow hairs on the inner surface of the petals.

Yellow is an unusual color for mariposa lilies, and there are only two other yellow-flowering species in our area. Impossible to miss and breathtakingly beautiful when seen in large numbers after chaparral fires is the yellow mariposa lily *(C. clavatus).* This tall species, restricted to the northern part of our region, has a zigzag stem topped by two to five large cup-shaped flowers up to two inches across (pl. 221, bottom left.

A number of other beautiful small geophytes are abundant but have less mass flowering. These species all differ from the mariposa lilies in having smaller flowers with sepals and petals that are alike in color and size, thereby giving the impression of six petals compared to the three present in the mariposa lilies. A very common and widespread species in early spring is blue dicks *(Dichelostemma pulchellum),* with its tall leafless stem topped by a tight cluster of blue purple flowers (pl. 222). Often growing with blue dicks in sites with heavy

Catalina mariposa lily (*Calochortus catalinae,* Liliaceae), April to May.

Lilac mariposa lily (*Calochortus splendens,* Liliaceae), May to June.

Butterfly mariposa lily (*Calochortus venustus,* Liliaceae), May to June.

Weed's mariposa lily (*Calochortus weedii,* Liliaceae), May to June.

Yellow mariposa lily (*Calochortus clavatus,* Liliaceae), April to June.

Plate 221. Mariposa lilies.

Plate 222. Blue dicks (*Dichelostemma pulchellum*, Liliaceae), April to June.

Plate 223. Harvest brodiaea (*Brodiaea jolonensis*, Liliaceae), April to May.

clay soils, but flowering later in the spring, are two close relatives that also produce their flowers at the tips of naked stems.
Harvest brodiaea *(Brodiaea jolonensis)* is a shorter plant with

Plate 224. Golden stars (*Bloomeria crocea*, Liliaceae), April to June.

Plate 225. Chocolate lily (*Fritillaria biflora*, Liliaceae), February to April.

inch-long violet blue tubular flowers arrayed in an open cluster (pl. 223). Golden stars *(Bloomeria crocea)* has tall stems like those of blue dicks, each topped by an open cluster of 30 to 50 yellow orange flowers (pl. 224). Each tiny petal has a prominent brown stripe down its center. Grassy fields filled with these plants in late spring twinkle in the late afternoon sun, giving rise to their common name.

Much less common but perhaps our most beautiful grassland geophyte is the chocolate lily *(Fritillaria biflora)*. Although this is the only species of the many fritillaries of California that occurs in our region, it has been termed the Cleopatra of the fritillaries because of its dark beauty. The large nodding chocolate brown flowers of this species easily distinguish it from any other plant in our region (pl. 225).

Plate 226. Blue-eyed grass (*Sisyrinchium bellum*, Iridaceae), March to June.

Chocolate lilies are uncommon primarily because they grow on heavy clay soils.

Two more grassland or open-site geophytes represent unusual distributions of their families in the foothills and coastal areas of Southern California. One of these is blue-eyed grass *(Sisyrinchium bellum)*, the only member of the iris family (Iridaceae) outside of higher-elevation forests. Rather than having a bulb or corm, this species, like many irises, has an underground stem or rhizome as a storage organ. The basal leaves of blue-eyed grass are grasslike and extend upward to a height that matches that of its flattened flowering

Plate 227. Rein orchid (*Piperia cooperi,* Orchidaceae), May to June.

stalks, which are topped by small blue flowers with yellow centers (pl. 226). The early Spanish settlers made a tea from the rhizome to treat fevers.

Rare but worth a search is the rein orchid *(Piperia cooperi).* This slender orchid with a tall spike of small greenish white flowers (pl. 227) can occasionally be found in dry grassy openings of chaparral and woodlands, with notable populations at Torrey Pines State Park. Only one other orchid is widespread in the coastal and foothill areas of Southern California, the stream orchid *(Epipactis gigantea)* described in chapter 8.

Riparian Environments

Drainage channels incised into hillsides by erosion over long periods of geologic time extensively cut the slopes of the many mountain ranges of Southern California. These small channels join into larger streams and rivers that eventually reach the coast. The flow of water through these stream channels allows the development of distinctive streamside communities that are termed riparian woodlands. Hikers in the foothills are usually quite familiar with these riparian habitats, particularly during the summer months, when the shade of tall sycamores *(Platanus racemosa)* and coast live oaks *(Quercus agrifolia)* provides a welcome respite from the sun (pl. 228). The riparian woodlands are only well developed along stream channels where water remains available throughout the year.

The fire and flood cycles and general geological instability

Plate 228. Live oaks and willows along Malibu Creek in the Santa Monica Mountains.

that characterize Southern California make riparian habitats highly dynamic environments. Any long-term resident of a mountain canyon can attest to the dramatic changes that may occur when fire denudes hillsides, promoting erosion and landslides when the winter rains arrive. Flooding resulting from heavy rain may deeply scour stream and creek channels in one year and allow the accumulation of thick layers of sand and debris in other years. Such events, influenced by a complex mixture of natural and human-influenced processes, make the riparian zone a constantly changing environment. It is not surprising, therefore, that the structure of riparian plant communities is complex and highly variable in both time and space.

Riparian habitats represent a gradient of landscape forms. These range from broad woodland corridors along perennial streams that drain large watersheds to small canyons fed by intermittently flowing streams. The more water that remains available along the riparian corridor during the dry summer months, the richer is the development of the riparian zone community and the greater its biological diversity. Some riparian tree species, such as white alder *(Alnus rhombifolia)* and Fremont and black cottonwood *(Populus fremontii* and *P. trichocarpa)*, are best developed on the wider riparian floodplains lower along perennial streams. Others, such as California or western sycamore, live oaks *(Quercus agrifolia* and *Q. chrysolepis)*, and California bay *(Umbellularia californica)*, occur much more widely and are found along the many intermittent stream channels throughout the mountains. Higher up in the mountains, where the streams and runoff channels become much narrower and steeper, water becomes much less available, and summer drought begins to limit the presence of riparian species.

Even though water may not be visible along many creek channels during the dry summer months, some subsurface flow may still be present. Riparian trees are usually a good indication of the presence of such water because they do not survive if soils become dry for even short periods. Trees such as sycamore, willows (*Salix* spp.), cottonwoods, and bigleaf maple *(Acer macrophyllum)* all require large amounts of water and only grow where ample water occurs near the surface throughout the year. When you see such species growing along steep creek channels high in the mountains, you can be certain that some semipermanent flow or perched pool of water is available to nurture them through dry summer conditions.

An interesting feature of many riparian tree species in Southern California is that they lose their leaves for several months during the winter, generally from December until March. In comparison, the coastal sage scrub species described in chapter 3 lose their leaves in the summer. Winter leaf fall indicates an ancestral relationship to tree species in cold climates like those of the eastern United States. By losing their leaves during cold winter months, trees recycle nutrients and save carbohydrates that would otherwise be necessary to maintain those leaves. The color changes in most deciduous species come about as the chlorophylls that give leaves their characteristic green color are broken down by enzymes and the associated nutrients are recycled for new leaves the following spring. With the chlorophylls gone, the remaining plant pigments become visible as fall colors. The limited fall colors in our mountains come from changes in the leaves of sycamore, willow, and California walnut *(Juglans californica)* in November and December.

Riparian woodlands cover only a very small portion of Southern California, but their ecological significance is far greater than their size. Situated as they are in the interface between the drier chaparral and oak woodland communities on the slopes of the mountains and the aquatic ecosystems of the streams themselves, riparian communities play a critical role in reducing erosion and stabilizing stream channels (pl. 229). The extensive root systems of riparian trees hold soils and stream banks in place and help flowing stream water to

Plate 229. California sycamores, which characterize areas with available subsurface water throughout the year.

recharge soil groundwater pools. Where riparian woodlands have been destroyed or reduced in area by human disturbance or severe fire damage, it is common to see dramatic increases in rates of storm runoff and erosion. When destruction of riparian plant cover is combined with loss of shrub cover on chaparral slopes above the streams, catastrophic mudflows can occur in years with heavy rains. Where the riparian community remains intact, however, such flows are greatly reduced, even when fires have denuded chaparral slopes.

Riparian vegetation plays an important role in regulating the movements of wildfires. Because moisture is readily available to riparian trees, the tissues of these species retain high water contents even under drought conditions. This characteristic makes these communities much less flammable than chaparral or other woodlands. Additionally, riparian areas are topographic low points that wildfires do not readily enter because heat plumes generally rise. Thus, only light ground fires typically enter riparian corridors, and these move slowly, without the high intensity of chaparral fires. As a result, riparian corridors are often effective barriers limiting the extent of fires.

Much of the terrestrial and aquatic wildlife in Southern California depends either directly or indirectly on riparian communities for survival. Riparian woodlands provide important structural complexity, with mosaics of cool, shady habitat for shelter in summer; protected wildlife corridors for movement of larger animals between adjacent plant communities; and year-round availability of water, food sources, and nutrient-rich organic sediments. It is not surprising, therefore, that riparian zones are centers of high biodiversity.

Riparian Trees

Most of the dominant trees in the riparian woodlands of Southern California are easily recognized. California sycamore, with its large five-lobed leaves and white bark, is one of the most characteristic (pl. 230). Although sycamores may occur in stream bottoms, they are more commonly rooted in floodplain deposits of soil slightly above stream channels. California sycamores can reach truly massive sizes. Large trees, often having multiple trunks, can attain heights of up to 100 feet or more and diameters of as much as five feet. Because these trees have relatively shallow roots, they only occur where water remains available at shallow soil depths throughout the year. Sycamores can often be seen at the heads of small seasonal streams on relatively steep slopes, but careful examination of these sites reveals the presence of small springs that maintain flows throughout the dry summer and fall.

Sycamore can be confused with only one other tree, bigleaf maple, but the large five-lobed leaves of bigleaf maple are much more deeply lobed (pl. 231). Bigleaf maple is widespread in shaded canyon bottoms and on moist canyon slopes with artesian springs, although it is much less abundant than

Plate 230. California sycamore (*Platanus racemosa*, Platanaceae), February to April.

Plate 231. Bigleaf maple (*Acer macrophyllum*, Aceraceae), April to May.

sycamore in Southern California. Stands of bigleaf maple are much more characteristic of the coastal forests of northern California and the Pacific Northwest.

Coast live oak *(Quercus agrifolia)* is one of the most abundant trees along riparian corridors (see pl. 157). The trees are seldom rooted in stream channels, instead becoming established on adjacent slopes where soils remain well drained. Rather than using stream water directly, these oaks typically tap permanent groundwater pools lying below the streams. Although the oaks are one of the most important components of riparian woodlands because of their abundance and strong evergreen cover that shades these habitats, they are not at all restricted to riparian areas. Coast live oaks can become established anywhere that their deep roots can reach permanent groundwater pools.

Another live oak species often occurs with coast live oak along streams at higher elevations above about 3,000 feet in the Transverse and Peninsular Ranges. This is canyon live oak *(Q. chrysolepis)*, a beautiful tree that grows to large size. Its evergreen leaves have a powdery gold color on their backs when young that turns blue gray as the leaves mature (pl. 232).

The willows, an important group of riparian tree species, may range from shrubby plants no more than six feet in height to large trees reaching to over 50 feet. Willows are abundant along perennial streams in Southern California and form broad areas of thickets on lower floodplains. All willows are characterized by easy resprouting of new stems from their underground roots, which grow directly in the bottoms of stream channels. This resprouting is an important adaptation for growth and survival in this dynamic environment, where periodic heavy flows of water may partially uproot smaller trees and scour or add new sediments to the stream channel. Rapid growth following disturbance is a notable characteristic of willows and inspired their Latin name, *Salix,* which means to leap. Willows can even form roots and become reestablished from broken trunks or branches. Given these characteristics, it is not surprising that willows are generally the woody species best able to colonize open and disturbed riparian habitats.

An abundance of windblown seeds also helps willows to

find and colonize new available stream habitats. Willows have two sexes. Half of the trees have only male flowers, which produce pollen, while the other half have female flowers, which produce seeds when pollinated. Because they are pollinated by wind, the flowers have no need to be showy to attract pollinators. Instead the small flowers of both male and female individuals are arranged in drooping clusters called catkins. These generally appear in early spring well before the growth of new leaves.

Many of our Southern California willows look very much alike and are difficult to separate at first. Once you are familiar with them, however, they will be much easier to identify. The most common is arroyo willow *(Salix lasiolepis)*, a common species along streams throughout California (pl. 233). Arroyo willow commonly reaches 10 to 20 feet in height. Also com-

Plate 232. Canyon live oak (*Quercus chrysolepis,* Fagaceae), April to May.

Plate 233. Arroyo willow (*Salix lasiolepis,* Salicaceae), February to April.

Plate 234. Red willow (*Salix laevigata,* Salicaceae), March to May.

mon is red willow *(S. laevigata)*, sometimes called black willow, which likewise forms large trees along streams throughout the state. Red willow is generally larger than arroyo willow and is characterized by leaves that are broadest near the base and taper steadily toward the tip, and by black bark (pl. 234). The smaller arroyo willow has leaves that are widest midway along their length and gray bark.

The most distinctive of our Southern California willows is sandbar willow *(S. exigua)*, a small tree that is sometimes called narrow-leaf willow. Fine hairs cover both sides of the leaves, giving them a distinctly silky or silvery appearance (pl. 235). Unlike most *Salix* species, sandbar willow does not form catkins until after it forms new leaves in spring.

Willows of all species played an important role in the life of Native American tribes in Southern California. Willow stems were widely used for baskets and as flexible poles in the construction of houses. Willow bark was used to cure fevers

Plate 235. Sandbar willow (*Salix exigua,* Salicaceae), March to May.

and to make skirts for women. We know today that this bark contains salicylic acid, the chemical from which aspirin was first synthesized.

Two species of cottonwoods are among the largest riparian trees in Southern California. Like willows, cottonwoods are excellent colonizers of open riparian soils. Cottonwoods also share the trait of having separate male and female individuals. Female trees have an unbelievable ability to produce and widely disperse their small wind-borne seeds. A single mature female tree can produce 25 million seeds in a single year! These seeds readily germinate when they encounter moist open soils along river channels, and seedlings grow rapidly until their roots reach permanent water pools. Fremont cottonwood, the more common of the two species, may reach 75 feet in height. Its leaves are green on both upper and lower surfaces (pl. 236). Black cottonwood, whose leaves are silvery or whitish below, may reach over 100 feet in height and have a trunk two to three feet in diameter. Both of these species are largely confined to broad riparian zones at lower elevations.

White alder is locally abundant in Southern California,

Plate 236. Fremont cottonwood (*Populus fremontii,* Salicaceae), March to April.

where it forms thickets of small to moderate-sized trees on floodplains along permanent streams. White alder is much more characteristic of riparian woodlands in central and northern California, where it may form dense stands extending for miles along stream channels. Its deciduous leaves, ovate in shape and two-and-a-half to three inches in length, are characterized by prominent parallel veins (pl. 237). Unlike all our other deciduous species, white alder drops its leaves while they are still fully green. This is because alders have special root nodules containing microorganisms called actinomycetes that fix nitrogen directly from the atmosphere. With such a ready supply of nitrogen, alders apparently do not need to break down and recycle nitrogen from the chlorophylls of their leaves before shedding them.

California bay is unusual among the riparian trees of Southern California (excluding the live oaks that are not strictly riparian) because of its evergreen leaves (pl. 238). The pungent odor of its leaves makes this species easy to recognize. California bay commonly occurs as isolated trees up to

Plate 237. White alder (*Alnus rhombifolia*, Betulaceae), January to April.

Plate 238. California bay (*Umbellularia californica*, Lauraceae), December to May.

30 feet in height along stream banks, usually well above the stream channels, or on shaded moist slopes well above streams where there are springs or other local sources of soil moisture throughout the dry summer months. At times, however, California bay may form almost pure stands where springs supply a steady source of water throughout the year. Seedlings and saplings are quite tolerant of shade. California bay is the only common riparian tree species that is insect pollinated.

As a member of the avocado family (Lauraceae), California bay is closely related to the European laurel *(Laurus nobilis)*, the traditional source of bay leaves for cooking. California bay, however, often provides the bay leaves sold in California supermarkets today. A handful of bay leaves collected and air dried are useful in any kitchen. Because the odor and flavors of California bay are somewhat stronger than those of European laurel, a single leaf is sufficient to flavor recipes.

California walnut forms a multibranched small tree along small creeks in shaded canyons in many parts of Southern California. Although sometimes present along the edges of riparian woodlands, walnut is more characteristic of open

Plate 239. Arizona ash (*Fraxinus velutina,* Oleaceae), March to April.

woodlands well away from riparian areas. This species is discussed in chapter 6.

Our last riparian tree in Southern California is Arizona ash *(Fraxinus velutina)*, which may form a moderate-sized tree up to 30 feet in height. This deciduous tree is characterized by leaves that are made up of five distinct leaflets and borne in opposite pairs on the branches (pl. 239). Arizona ash is not abundant and is largely restricted to shaded streamsides.

Other Riparian Plants

Although many hikers never look beyond common tree species in exploring riparian woodlands, many shrubby and herbaceous species are likewise restricted to these habitats. Even though riparian habitats are relatively small in total area, they are critically important contributors to the biodiversity of our region. Studies in the Santa Monica Mountains have suggested that as many as 20 percent of all plant species have riparian woodlands or associated moist canyon bottoms as their typical habitat. This is a remarkable number, considering that these riparian habitats make up less than one percent of the land area.

Where typical riparian tree species are absent, large riparian shrubs can form their own dense canopies. One common riparian shrub species in open areas along streams is mule fat *(Baccharis salicifolia)*, a member of the sunflower family (Asteraceae). Older books referred to this species as *B. viminea* or *B. glutinosa.* The latter name is appropriately descriptive, as the leaves of mule fat are distinctly sticky to the touch. Mule fat is widespread along both permanent and seasonal streams, where it may form large thickets up to 12 feet in height. Mule fat has a very willowlike appearance, with multiple vertical branches and narrow evergreen leaves two to four inches in length (pl. 240). A related species of large shrub is broom baccharis *(B. sarothroides)*, which is common in San

Plate 240. Mule fat (*Baccharis salicifolia*, Asteraceae), January to December.

Plate 241. Broom baccharis (*Baccharis sarothroides*, Asteraceae), June to October.

Diego County and extends out into desert washes to the east (pl. 241).

The great majority of shrub and herb species in riparian woodlands grow in shade or semishade beneath the canopies

Plate 242. Mugwort (*Artemisia douglasiana*, Asteraceae), July to November.

of the large trees. Most of these species grow not in the streambed itself, where they could be easily swept away in floods, but rather on the stream banks. One of the most widespread of these is mugwort *(Artemisia douglasiana)*, another member of the sunflower family, which reaches three to five feet in height. It can also be encountered in moist hollows within live oak woodlands. Mugwort reproduces well from rhizomes and thus is often found in large clonal stands. It is a very easy shrub to recognize because of the gray green stems and foliage. The large leaves are pale green and hairless above but gray and densely woolly below and deeply toothed at the tip (pl. 242). Tradition says that leaves of mugwort rubbed over the skin can help prevent rashes from forming after exposure to poison oak. Although it may well have some basis, this tradition is not well tested scientifically, so avoidance of poison oak is the preferred strategy. Many consider a poultice made from mugwort leaves an effective remedy for insect bites or stinging nettle.

Two native trailing shrub species often form tangled masses of stems along stream banks in riparian woodlands. A

Plate 243. California blackberry (*Rubus ursinus*, Rosaceae), February to June.

familiar one is California blackberry *(Rubus ursinus)*. The genus name *Rubus* for blackberries comes from the Latin word for bramble. The spiny stems of California blackberry trail for up to 20 feet and build mounds that serve as favored habitats for many small mammals and birds. Although the three leaflets and trailing growth form of blackberries lead many people to confuse these plants with poison oak, the saw-toothed leaf margins and spiny stems readily distinguish blackberries (pl. 243). They are most characteristic of stream banks but can be found anywhere there are seepage zones or available water. The berries are delicious and were widely used by Native Americans as a food source in summer. Collecting them in large amounts often presents a challenge, however, because of their small size and the prickly stems. This native species is the primary ancestor of cultivated boysenberries and loganberries.

A related species and one that shares the thorns and growth form of the California blackberry is California wild rose *(Rosa californica)*. Wild rose grows well along shaded stream banks but also thrives in open habitats where seepage water is available. Rose patches are commonly three to four feet high, but old individuals occasionally reach double that height. Although it may be confused with blackberry at first encounter, California wild rose is easy to distinguish by the

presence of five to seven leaflets rather than the three of California blackberry and by its pink flowers, which are larger than the small white flowers of the blackberry (pl. 244).

Plate 244. California wild rose (*Rosa californica,* Rosaceae), April to July.

Southern California is also home to an interesting vine, desert wild grape *(Vitis girdiana),* that grows luxuriantly in canyon bottoms and along streams, often forming masses of stems that drape over large trees (pl. 245). This species has an unusual pattern of distribution, occurring in scattered sites throughout

Plate 245. Desert wild grape (*Vitis girdiana,* Vitaceae), May to June.

Plate 246. Skunkbrush (*Rhus trilobata,* Anacardiaceae), March to April.

Southern California and again in the desert mountain ranges of Inyo County.

Skunkbrush *(Rhus trilobata)*, once known as squaw bush and now named for the odor of its leaves, is a widespread but infrequent shrub growing up to five feet in height. It occurs commonly in canyon bottoms beneath the semishade of sycamores or oaks. The three leaflets (pl. 246) closely resemble those of poison oak, and indeed the two species have some distant relationship. However, skunkbrush lacks the toxic chemicals of poison oak and presents no problem if it is touched. One way to distinguish the two, beyond the general shrubbier growth form of skunkbrush, is to look at whether the terminal leaflet is stalked or merges directly into the base of the other two leaflets. Poison oak has a stalked terminal leaflet, whereas squaw bush does not. In summer, skunkbrush has red berries, and those of poison oak are white. Both species lose their leaves in winter. Split stems of skunkbrush were widely used by Native American women in wrapping the coils of their baskets. The berries were collected and steeped in boiling water to make a beverage. Powdered berries were used to treat a number of conditions, including smallpox.

Less common and unusual in many respects is leather root *(Hoita macrostachya,* formerly known as *Psoralea macrostachya).* This large shrub, a member of the pea family (Fabaceae), may reach eight to 10 feet in height, but the stems are not strongly woody and are easily broken. Leather root is simple to recognize, with its large, pungent trifoliate leaves (pl. 247). In summer leather root produces showy spikes four inches in length with purple flowers resembling those of the pea. This genus is the subject of widespread medical interest because it produces a chemical compound used in the treatment of certain cancers, including T-cell lymphomas in AIDS patients.

Plate 247. Leather root (*Hoita macrostachya,* Fabaceae), May to October.

Plate 248. Giant chain fern (*Woodwardia fimbriata,* Blechnaceae).

Although ferns are not a diverse part of our flora in Southern California, stream banks and canyon walls are excellent habitats for fern growth. With a little luck, exploration of cool, well-shaded canyons with perennial streams will reveal

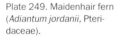

Plate 249. Maidenhair fern (*Adiantum jordanii,* Pteridaceae).

individuals of the giant chain fern *(Woodwardia fimbriata)*. This large fern, more typical of the redwood forests of northern California, has masses of fronds reaching six to eight feet in length (pl. 248). On the rock walls of these canyons, particularly in areas of seepage, you may find delicate stalks of maidenhair fern *(Adiantum jordanii)*, with its fan-shaped leaves (pl. 249). The stream banks of these canyons are often home to goldback fern *(Pentagramma triangularis)* (pl. 250) and California polypody *(Polypodium californicum)* (pl. 251).

There is insufficient space here to describe and illustrate the large number of herbaceous species associated with riparian habitats in Southern California. It would be impossible, however, not to include three of the most spectacular of these. The first is the Humboldt lily *(Lilium humboldtii)*, also called leopard lily, which is common along the banks of intermittent

Plate 250. Goldback fern (*Pentagramma triangularis,* Pteridaceae).

Plate 251. California polypody (*Polypodium californicum,* Polypodiaceae).

Plate 252. Humboldt lily (*Lilium humboldtii* subsp. *ocellatum,* Liliaceae), May to June.

Plate 253. Stream orchid (*Epipactis gigantea,* Orchidaceae), April to May.

Plate 254. Scarlet monkeyflower (*Mimulus cardinalis,* Scrophulariaceae), May to October.

streams in the shade or partial shade of oaks. This beautiful species produces groups of large nodding flowers at the top of a stout stem that may be up to five to eight feet or more in height (pl. 252). The linear leaves are arranged in discrete whorls around the stem. Much less common but worth a search along shaded perennial streams is the stream orchid *(Epipactis gigantea)*. This is a tall species up to three to four feet in height, with solitary flowers formed in each of the leaf axils along the stems (pl. 253). Last but not least is scarlet monkeyflower *(Mimulus cardinalis)*, a stout perennial herb two to four feet in height (pl. 254). This is the only red-flowered species of monkeyflower in our flora.

Estuaries and Salt Marshes

Estuaries are transition zones along the coast where rivers meet the ocean. Rapidly fluctuating water levels and daily shifts in salinity make these environments harsh for most plant species. Nevertheless, reduced competition allows a few strong species to be spectacularly successful. With mild climate conditions and constant availability of water and nutrients in sediment flow, estuaries are incredibly productive ecosystems, even more productive per unit area than the tropical rainforests of the world. The carbon from their living plants and dead organic mass is the basis for food chains stretching from algae to invertebrates to vertebrate fish and birds, not only in the estuaries themselves but also in the surrounding marine environment. An estimated 85 percent of all the fish and shellfish now sold in commercial markets of the world spend all or part of their lives in estuaries. Despite the critical ecological importance of estuaries and associated salt marshes, an estimated 75 to 90 percent of these wetlands in Southern California have been lost to dredging or filling in the last century. Much of the small remaining area has been heavily degraded.

There are several forms of estuaries and associated salt marshes in Southern California, associated with differences in local geology and hydrology. The largest salt marshes are found in bay estuaries where gentle topography provides an open connection with the ocean. Morro Bay, Bolsa Chica, Upper Newport Bay, Mission Bay, San Diego Bay, and the Tijuana Slough are all examples of bay estuaries with extensive salt marshes (pl. 255). This form of estuary is rare along the Santa Monica Mountains in Los Angeles County and the coasts of Ventura and Santa Barbara Counties where coastal uplift is rapid. Estuaries along these coasts are small and their associated salt marshes much less extensive. The Santa Clara and Ventura Rivers in Ventura County furnish examples of river mouth estuaries, where perennial river flows provide permanently brackish conditions without the formation of extensive mudflats. At the mouth of Malibu Creek in Los Angeles County and in many of the smaller estuaries in San Diego County, stream flow is reduced enough in summer to allow sandbars to build up and prevent outward flow, forming lagoons. Goleta Slough and the Carpenteria Salt Marsh in Santa Barbara County are structural basins where geologic faulting has created narrow estuaries.

Estuary habitats range from subtidal seagrass communities with eel-grass *(Zostera marina)* and surf-grass *(Phyllospadix torreyi)* to salt marshes on the shoreward margin of mudflats. Because the theme of this book is terrestrial communities, we will focus on salt marshes. One of the first things to notice about these habitats is that there are relatively few dominant plant species. Moreover, there is commonly a striking zonation of plant distributions correlated with the level of tidal flow. One group of species covers much of the area where water sits on the soil surface for most of the day. With just a

Plate 255. The Bolsa Chica salt marsh in Orange County with pickle-weed, saltwort, salt grass, cord grass, and sea lavender as dominant species.

small increase in elevation, soils are alternately submerged and exposed with tidal changes, and a different plant community thrives. Areas at the edge of the marsh, where only the highest tides can reach, are different as well.

The diversity of plants in salt marshes is limited by the salinity levels of the soil. Plants able to survive and grow under saline conditions are termed halophytes. High internal levels of sodium and chloride, the primary components of ocean salt, are toxic to most plants. Plants must use one of two primary physiological modes of adaptation to survive salt marsh conditions. Either they must exclude potentially toxic salt when they take up water through their roots, or they must be able to tolerate high levels of salt in their tissues. The former strategy requires large amounts of metabolic energy, whereas the latter presents the problem of keeping salts from building up to toxic levels in leaves. Most of our salt marsh plants accumulate salts as they take up water, but they utilize different strategies to eliminate these salts.

Let's start with plants that adapt to saline soils by tolerating high salt concentrations in their tissues. The best way to deal with increasing levels of tissue salt is to store it away in a form that will not disrupt the metabolism of the plant cells. This can be accomplished by concentrating salts in membrane-lined cell structures called vacuoles, isolated from cytoplasm functions of the cell. The continuing need to store more salts requires the cells of such plants to be large, so their tissues are noticeably succulent. Eventually, however, individual cells reach their storage capacity. The plants then shed salt-laden leaves or stems and grow new young tissues.

One of our most widespread and characteristic plant groups in the salt marshes of Southern California, the pickleweeds, illustrates this mechanism. The peculiar fleshy jointed stems of pickleweed mass together to form dense low-growing shrubby mats along the middle tidal levels. These jointed stems and oppositely arranged side branches resemble small pickles. The common name may also come from a long tradition in Europe of eating young succulent branches as a fresh or pickled vegetable. Another common name once used was glasswort, because these plants were once harvested by the glass industry along European coasts for the high content of soda in the tissues. Southern California has two common species of pickleweed, *Salicornia virginica* (pl. 256) and *S. subterminalis* (pl. 257). By late summer, pickleweed typically turns from spring green to a reddish green or even deep rose color, forming a striking element of salt marsh plant cover.

Plate 256. Pickleweed (*Salicornia virginica*, Chenopodiaceae), April to September.

Plate 257. Salt marsh bank dominated by pickleweed (*Salicornia subterminalis*, Chenopodiaceae), April to September.

Two other common salt marsh plants in Southern California also illustrate the use of succulent plant tissues to hold salt. The more striking of these, in a community where showy flowers are rare, is marsh jaumea *(Jaumea carnosa)*. This fleshy perennial grows within the pickleweed zone with stems six to 10 inches in length trailing along the soil surface and rooting at their nodes (pl. 258). Marsh jaumea is a member of the sunflower family (Asteraceae) and has small heads of yellow flowers. The other common fleshy species, saltwort *(Batis maritima)*, is much less showy. Saltwort is a sprawling perennial herb, with stems trailing from a woody base for three feet or more along the soil surface. The stems bear small lateral branches with oppositely arranged fleshy leaves (pl. 259). Like those of pickleweed, the flowers are small and easy to overlook.

A second group of common salt marsh plants tolerates salt within cytoplasm but uses specialized structures called salt glands to excrete excess salt. The most characteristic of these species is salt grass *(Distichlis spicata)*, a trailing perennial grass encountered in stiff mats at the upper margins of salt

Plate 258. Marsh jaumea (*Jaumea carnosa,* Asteraceae), June to October.

Plate 259. Saltwort (*Batis maritima*, Batidaceae), July to October.

marshes, commonly just above the mean high-tide line (pl. 260). The salt glands of salt grass are located in rows parallel to the veins on both upper and lower leaf surfaces. The salt gland structure includes a basal cell where salts collect and a cap cell that connects with the outer leaf surface and is the site of salt excretion. Looking closely at the leaves of salt grass,

Plate 260. Salt grass (*Distichlis spicata*, Poaceae), April to July.

particularly with a hand lens, it is easy to see the salt crystals lying on the leaf surface above each gland. Lacking a hand lens, the presence of these crystals can be readily tasted.

Another salt-excreting perennial grass is cord grass *(Spartina foliosa)*, which dominates mudflats at the lowest zone in salt marshes (below the middle pickleweed belt). The tall stems of cord grass (pl. 261) are covered by water at high tides, giving them the appearance of rice paddy plants, and are partially submerged for the majority of the day. The high productivity of cord grass and the large amount of dead organic matter that it generates each year make this a keystone species in salt marsh food chains. Although cord grass forms extensive pure stands at the lower margins of bay estuaries in San Diego County and in Upper Newport Bay, as well as along the central and northern California coasts, it is rare or absent from the smaller salt marshes in Los Angeles, Ventura, and Santa Barbara Counties.

Two final species of common salt marsh plants with evident salt-secreting glands should be mentioned. Sea lavender *(Limonium californicum)*, an herbaceous perennial with a

Plate 261. Cord grass (*Spartina foliosa,* Poaceae), July to November.

Plate 262. Sea lavender (*Limonium californicum,* Plumbaginaceae), July to December.

Plate 263. Alkali heath (*Frankenia salina,* Frankeniaceae), June to October.

woody base, typically grows in the pickleweed zone. It is easy to recognize with its basal clump of large oblong leaves (pl. 262). The small purple flowers are borne on upright stems in summer but are inconspicuous and easy to miss. European species of sea lavender have larger flowers and are widely cultivated for use in dried flower arrangements. Alkali heath *(Frankenia salina)* is common in the upper salt marsh zones, often mixed with salt grass. This herbaceous perennial reaches up to a foot or more in height, with small oppositely arrayed leaves displaying salt crystal residues on their surface (pl. 263). Like marsh jaumea, alkali heath is one of the few salt marsh plants with showy flowers pollinated by insects. The attractive pink flowers provide color in summer to the otherwise bland salt marsh flora.

Freshwater Marshes

Freshwater marshes are plant communities that occur in areas where standing or slowly flowing water saturates or floods the soils throughout the year (pl. 264). Freshwater wetlands probably were never very common in Southern California, because most areas saturated in winter are dry during summer. They were, however, much more common than they are today. These wetlands would have been found in permanently saturated soils along low-lying bottomlands bordering perennial streams and in hollows behind the crests of coastal dunes where pools of freshwater float over saltwater below. Now that the major rivers flowing across Southern California have been channelized in concrete flumes, most of the bottomland marshes have disappeared. Extensive areas of dune hollow marshes once occurred along the Oxnard Plain and along the coast of Los Angeles from Playa Vista to Redondo Beach. Urbanization of the coast over the past century, how-

Plate 264. A freshwater marsh with narrow-leaved cattails *(Typha domingensis)* in the background and evening primrose *(Oenothera elata* var. *hookeri)* in the foreground, Santa Monica Mountains.

ever, has eliminated most of these marshlands. Examples of dune marshlands can still be seen today at Vandenberg Air Force Base and around the Nipomo and Guadalupe dunes of the central coast.

One of the most characteristic and easily recognized large freshwater marsh plants is the cattail, familiar to most schoolchildren because of its fuzzy brown flower spikes (pl. 265).

Plate 265. Broad-leaved cattail (*Typha latifolia*, Typhaceae), June to July.

The long narrow leaves of cattails grow in a sheath around the flowering stem, which reaches 4 to 9 feet or more in height. The narrow upper portion of the flowering spike consists entirely of male flowers, which produce copious amounts of pollen. Female flowers form the lower and thicker half of the spike. At the end of the summer these female flowers begin to break apart from the stems and disperse thousands of puffy plumed seeds. Cattails were an important plant for California Indians. They used the juvenile upper spikes of the flowering stalk as a vegetable and baked the fleshy rhizomes as a potato-like food or dried and ground them to make flour. Cattail leaves were woven into mats or baskets, or used as thatch for shelters. Southern California has two common species of cattail. The more common is the broad-leaved cattail *(Typha latifolia)*, distinguished by the lack of any separation on the flowering stalk between the upper male and the lower female flowers. In the narrow-

Plate 266. Bulrush (*Scirpus californicus,* Cyperaceae), June to September.

leaved cattail *(T. domingensis),* there is a separation of about three-quarters of an inch between these groups.

A second group of important marsh plants that rise well above the water in dense stands is the bulrushes. These tall grasslike plants are members of the sedge family (Cyperaceae). Grass stems are round in cross section and hollow except at the nodes that occur along the stems. Sedges, in contrast, have solid stems (except for open chambers, as described below) that are usually triangular in cross section. From this characteristic comes the mnemonic "sedges have edges." The bulrushes are the largest sedges in California. The freshwater marshes of Southern California contain many species of bulrushes (pl. 266). The common tall species that reach eight to 12 feet in height are *Scirpus acutus* and *S. californicus.* These species are difficult to separate, but the stem of the former is oval in cross section, whereas that of the latter is triangular. Like those of cattails, the fleshy rhizomes of bulrushes can be dried and ground into flour. The seeds of bulrushes are an extremely important food for migratory waterbirds. Three other common groups of sedge family species are the true sedges *(Carex),* the nut sedges *(Cyperus),* and the spike rushes *(Eleocharis).* The freshwater marshes of our region contain many species in these groups.

Flooded marsh soils hold little oxygen because of the high demand for this gas by soil bacteria and small animals. Bacterial formation of a gas called hydrogen sulfide gives these anoxic soils a strong and unpleasant odor of sulfur or rotten eggs. Marsh plants have developed special means to survive in these soils. Both cattails and bulrushes have open air-filled chambers that interconnect down the stem and supply necessary oxygen to the roots. The California Indians took advantage of the large air spaces in bulrushes to make buoyant rafts of large clusters of bulrush stems lashed together with rope made from braided leaves of cattails.

The dense thickets formed by cattails and bulrushes in shallow marsh areas allow little room for other plant species to grow. Where water levels are deep, however, a number of smaller aquatic plants can be seen. The most widespread and showy of these is water smartweed *(Polygonum amphibium)*. It may float with its leaves on the water surface in deeper pools. More commonly it is a terrestrial plant at the muddy edges of ponds with upright leaves up to four feet tall and long spikes of rose-colored flowers arising in summer (pl. 267). Several related species of smartweed occur in our wetlands as well.

Plate 267. Water smartweed (*Polygonum amphibium* var. *emersum*, Polygonaceae), July to October.

Saturated soils of drainage channels and marshy areas often are covered by extensive stands of yerba mansa *(Anemopsis californica)*. This hollow-stemmed herb has basal leaves on long stalks reaching two feet in height and erect conical heads of flowering spikes with showy white bracts (petal-like specialized leaves) at the base (pl. 268). The creeping rootstalks of yerba mansa have a peppery, astringent flavor and were dried and ground by early Californians to brew a tea used as a blood purifier and a poultice for rheumatism.

Freshwater wetlands with open water are relatively uncommon in Southern California, but where they occur they may support abundant populations of water ferns *(Azolla filiculoides)*. These tiny mosslike floating plants have a worldwide distribution (pl. 269). They are widely used in rice paddies to enrich the soil because nitrogen-fixing algae are associated with their leaves.

Plate 268. Yerba mansa (*Anemopsis californica,* Saururaceae), April to July.

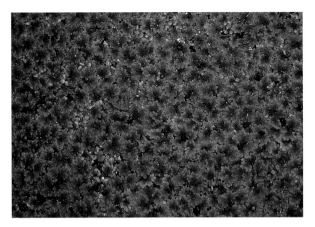

Plate 269. Water fern (*Azolla filiculoides,* Azollaceae).

Vernal Pools

Vernal pools, for which California is famous, are seasonal wetlands that form in the winter as rains fill shallow, poorly drained depressions. Rather than forming extensive landscape features, vernal pools in our area commonly exist as scattered small seasonal ponds. In spring these vernal pools explode with an amazing display of tiny flowers representing a diversity of species (pl. 270). By summer, however, the pools dry, with the vegetation turning completely brown (pl. 271). These are unusual habitats with geologic histories dating back millions of years. They are home to groups of rare plants found nowhere else in the world, as well as to rare crustaceans and other invertebrates whose ecology ties them closely to this specialized habitat.

Vernal pools in Southern California occur in scattered local areas where the combination of flat terrain and an impervious subsurface provides the necessary conditions for them to form each winter. The conditions are best developed

Plate 270. Vernal pool on the Santa Rosa Plateau in spring.

Plate 271. The same vernal pool as in plate 270 at the end of summer.

on coastal mesas such as Kearney, Miramar, and Otay Mesas in San Diego County, and on inland mesas such as those of the Santa Rosa Plateau in Riverside County. Small vernal pools once occurred in inland valleys of San Diego County around San Marcos, Ramona, and Temecula and in the Los

Angeles Basin, but these sites have been largely lost to urbanization. Likewise, vernal pools once occurred in scattered numbers along the margins of coastal bluff and dune areas of Los Angeles and Orange Counties, but none of these habitats remains today. It has been estimated that more than 97 percent of the original area of vernal pools in Southern California has been destroyed or heavily degraded.

The rich floras of vernal pools include two major groups of species. One group consists of widespread aquatic plants with worldwide ranges. The other group is made up of specialized species restricted to vernal pools and known only from California. The range of many of these latter species is restricted to a few local vernal pools. Although as many as 100 species are commonly associated with vernal pool habitats around California, no more than 15 to 25 plant species are typically present at any single pool. Subtle differences in the duration and pattern of ponding, soil characteristics, and water chemistry determine the species present at any single site.

Plants have evolved specialized traits to survive the sharp environmental changes that occur in vernal pool habitats over an annual cycle. Plants must endure aquatic conditions in winter when the pools fill, then act as dryland species as the pools begin to lose water in spring, and finally endure bone-dry soils in summer, when water is totally absent. If we look at the nearest relatives of vernal pool specialists, we can begin to understand how these adaptive traits have evolved. Vernal pool plants typically have their closest relatives among dryland, not wetland, species. This fact suggests that survival of summer conditions is a serious challenge. Most vernal pool specialists are annual plants that complete their life cycle from seed germination to death in a single year. Seeds provide an effective means of enduring extreme summer drought.

Some of the most interesting and showy vernal pool plants are diminutive annual species. The genus *Pogogyne,* in the mint family (Lamiaceae), includes nine species in California, all restricted to vernal pools. Two of these species occur in our region, including the San Diego mesa mint *(P. abramsii),* found in scattered pools on the Kearney and Miramar Mesas of San Diego County (pl. 272). Likewise tiny but showy are the downingias, including the widespread *Downingia cuspidata* (pl. 273). Masses of the multicolored blue, white, and gold flowers of these species are common in vernal pools in spring.

Not all vernal pool specialists have showy flowers. Two widespread examples that do not are San Diego button celery *(Eryngium aristulatum* subsp. *parishii),* a slender annual with tiny white flowers (pl. 274), and woolly marbles (*Psilocarpus* spp.), tiny, nondescript annuals (pl. 275).

Plate 272. San Diego mesa mint (*Pogogyne abramsii,* Lamiaceae), April to June.

Plate 273. Downingia (*Downingia cuspidata*, Campanulaceae), March to June.

Plate 274. San Diego button celery (*Eryngium aristulatum* subsp. *parishii*, Apiaceae), May to August.

Two grasslike species of vernal pool specialists have remarkable physiological strategies of adaptation. One of these is a true grass, the inconspicuous California Orcutt grass (*Orcuttia californica*). This species (pl. 276) and a closely related

Plate 275. Woolly marbles (*Psilocarpus brevissimus*, Asteraceae), April to June.

Plate 276. California Orcutt grass (*Orcuttia californica*, Poaceae), May to June.

Plate 277. Quillwort (*Isoetes howellii*, Isoetaceae) (photograph by Jon Keeley).

group of grasses known only from vernal pools in California and Baja California begin their growth in late spring when temperatures are high and most other vernal pool species have died back for the year. Careful investigations have shown that these grasses utilize C_4 metabolism, an unusual physiological system most characteristic of tropical perennial grasses adapted to high-temperature growth conditions. No other vernal pool species have this ability.

The other unusual adaptations are those of the quillworts (*Isoetes* spp.), a strange and ancient lineage of aquatic fern relatives that appear grasslike at first glance (pl. 277). These plants grow fully submerged and take in carbon dioxide from the water around them. Chemical changes in the acidity of small pools and lakes over the course of the day make dissolved carbon dioxide available at night but highly limited in the middle of the day. The quillworts have evolved a mechanism termed Crassulacean acid metabolism, more typical of cacti and other desert succulents, which allows them to fix carbon at night and store it for use in photosynthesis the following day.

Geography and Floristics

The Channel Islands comprise eight major islands located off the coast of Southern California (map 3). A northern group, an extension of the Santa Monica Mountains, includes the relatively large islands of Santa Cruz and Santa Rosa, and the smaller Anacapa and San Miguel. A southern group also includes two large islands, Santa Catalina and San Clemente, and the smaller San Nicolas and Santa Barbara. All of the islands are relatively low, with elevations above 1,000 feet reached only on the four largest islands, and a maximum elevation of 2,470 feet on Santa Cruz. The climate regime for all of the islands is maritime, with moderated temperature extremes and strong winds. The rainfall gradient from north to south is pronounced, with about 20 inches average annual rainfall on Santa Cruz and only about 11 inches on San Clemente.

Sea-level fluctuations through the glacial periods of the last million years had major effects on the size and shape of the islands. The four smaller islands were likely submerged during the peak sea levels of interglacial periods, and all four of the northern islands were linked together in one landmass about 18,000 years ago at the last glacial maximum. There is no evidence at this time, however, that this group of islands has ever had a land bridge connection to mainland Southern California, although their distance to the mainland was small.

As you might expect, the larger islands have more topographic, geologic, and climatic diversity than the smaller islands and thus have a broader range of community types. All of the major plant communities described in the previous chapters are represented on the islands, but in somewhat modified forms and sometimes highly restricted distributions. Because of the complexity of habitats present, the Channel Islands represent a number of interesting biogeographic patterns. Foggy maritime habitats mark the southern

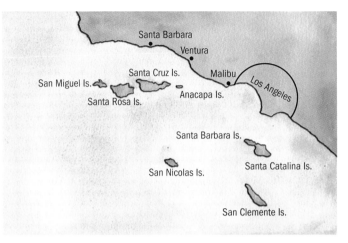

Map 3. The California Channel Islands (map by Lisa Pompelli).

limit of distribution for a number of species known only from central or northern California. The islands also contain species that the fossil record shows were once common on the mainland but that survive today only in favorable sites on the Channel Islands. In contrast, dry south-facing slopes of San Clemente and Santa Catalina mark the northern limits of distribution for a number of species characteristic of the coastal areas of northwestern Baja California.

There is a strong tendency for gigantism in both the growth form and the leaf size of the woody plants native to the Channel Islands. Shrubby species are taller and more treelike than mainland populations of the same and related species, and their leaves are generally larger. Moreover, many species that are herbaceous on the mainland have evolved more woody growth forms on the islands. Both tendencies are common on islands, representing evolutionary selection to fill unoccupied niches, and are reinforced by the maritime climate and relative rarity of fire on the Channel Islands.

Coastal Sage Scrub

Coastal sage scrub communities are well developed on all but the two smallest Channel Islands. They are typically found on dry rocky slopes and are most widespread on the south sides or south-facing slopes of the islands. As on the mainland, coastal sage scrub communities may be three to four feet in height, but on wind-exposed coastal sites they may be virtually prostrate (pl. 278). Many of the common shrub dominants are species discussed in chapter 3. These include California sagebrush *(Artemisia californica),* California encelia *(Encelia californica),* black sage *(Salvia mellifera),* lemonadeberry *(Rhus integrifolia),* and sawtooth goldenbush *(Hazardia squarrosa).* Added to these are three interesting endemic island species of buckwheat. Two of these, St. Catherine's lace or giant buckwheat *(Eriogonum giganteum),* present on San Clemente, Santa Catalina, and Santa Barbara (pl. 279), and Santa Cruz Island buckwheat *(E. arborescens),* present on all of the northern islands except San Miguel (pl. 280), are large shrubs reaching heights up to six feet or more. Giant buck-

Plate 278. Wind-trimmed coastal sage scrub on Santa Cruz Island (photograph by Mildred Mathias).

Plate 279. Giant buckwheat (*Eriogonum giganteum*, Polygonaceae), May to August.

Plate 280. Santa Cruz Island buckwheat (*Eriogonum arborescens*, Polygonaceae), April to September.

wheat has broad leathery leaves covered with white woolly hairs below, whereas Santa Cruz Island buckwheat has clusters of small linear leaves on the stems. Island buckwheat (*E. grande*), a spreading subshrub with a woody base support-

Plate 281. Island buckwheat (*Eriogonum grande*, Polygonaceae), June to October.

ing a rosette of large leaves (pl. 281), can be found on almost all of the islands. Growing with the coastal sage shrubs are the two prickly-pear species described earlier, coast prickly-pear *(Opuntia littoralis)* and tall prickly-pear *(O. oricola)*, and a variety of dudleya species.

Several additional species are present in coastal sage scrub communities growing on less arid north-facing slopes. These include golden yarrow *(Eriophyllum confertifolium)* and rockrose *(Helianthemum scoparium)*, described in chapter 3. Also common is giant coreopsis *(Coreopsis gigantea)*, a drought-deciduous shrub with a thick trunk reaching four to eight feet in height (pls. 282, 283). The trunk has only a few branches; the leaves form clusters and are finely divided into narrow linear lobes. The large yellow flower heads are up to three inches across and are displayed in clusters on long leafless stems. Although largely restricted to the northern Channel Islands, giant coreopsis is also present in scattered localities that re-

Plate 282. Giant coreopsis (*Coreopsis gigantea*, Asteraceae), March to May.

Plate 283. Giant coreopsis in its summer dormant phase on Anacapa Island.

ceive abundant fogs on the western end of the Santa Monica Mountains and along the coast north of Point Conception.

Heavily grazed areas of coastal sage scrub on Santa Cruz and Santa Rosa have had much of their shrubby diversity replaced by open stands of coyote brush *(Baccharis pilularis)*. It is not clear whether the coyote brush is the most resilient of the preexisting shrub species or is invading secondary grasslands dominated by European annual grasses.

Maritime Succulent Scrub

Maritime succulent scrub is restricted to the two large southern Channel Islands, San Clemente and Santa Catalina, where it is present on windy south-facing beach terraces. The dominant species have all been mentioned previously. These are coast prickly-pear, coast cholla *(Opuntia prolifera),* California boxthorn *(Lycium californicum),* and golden cereus *(Bergerocactus emoryi).* A notable absence is coastal agave *(Agave shawii),* a species common in similar mainland communities of maritime succulent scrub.

An interesting endemic species of this community is the showy island snapdragon *(Galvezia speciosa),* whose long tubular red flowers are very attractive to hummingbirds (pl. 284). This mound-forming shrub is known only from coastal rocky slopes and terraces on San Clemente, Santa Catalina, and Santa Barbara Islands in our region, as well as from Guadalupe Island off the central coast of Baja California. It is now rare over many parts of its original range as a result of grazing by feral animals.

Plate 284. Island snapdragon *(Galvezia speciosa,* Scrophulariaceae), February to May.

Island Chaparral

Island chaparral is most common on north-facing slopes of the Channel Islands where soils are rocky or shallow, but its occurrence is patchy today. Island chaparral is taller than mainland chaparral communities, with typical canopy heights of six to 12 feet (pl. 285). The general growth form of the domi-

Plate 285. Chaparral on Catalina Island with giant buckwheat, island redberry, toyon, crossosoma, and other species.

nant shrubs is also more arboreal, with more of a tendency toward a main trunk and a tall canopy. Well-developed stands of island chaparral are best observed on three of the larger islands—Santa Cruz, Santa Rosa, and Santa Catalina—as drier conditions restrict the extent of this community. Individual species of chaparral shrubs, however, are present in small numbers on some of the other islands.

Many of the common shrub species of this community either are the same as those on the mainland or are distinctive subspecies differing slightly in morphology. Chamise *(Adenostoma fasciculatum)*, toyon *(Heteromeles arbutifolia)*, laurel sumac *(Malosma laurina)*, lemonadeberry, and mission manzanita *(Xylococcus bicolor)* are all common shrub species in island chaparral that were described in chapter 4. California mountain mahogany *(Cercocarpus betuloides)*, bigpod ceanothus *(Ceanothus megacarpus)*, bush poppy *(Dendromecon rigida)*, and hollyleaf cherry *(Prunus ilicifolia)* have distinctive island subspecies. Island races of hollyleaf cherry, more correctly termed Catalina cherry, grow taller and more attractively than its mainland form and have found wide use in landscaping as hedges and ornamental plantings.

Island chaparral also includes a number of endemic shrub species, many restricted to only one or a few islands. Two endemics can be found on Santa Cruz, Santa Rosa, and Santa Catalina. These are island ceanothus *(Ceanothus arboreus)*, also known as feltleaf ceanothus, and island scrub oak *(Quercus pacifica)*, both large evergreen shrubs. Island ceanothus has broad, thin, dull green leaves that show three main veins and are covered with tiny hairs (pl. 286). Island scrub oak (pl. 287) is often considered to be only a form of mainland scrub oak, but researchers have now decided that it is a distinct species. Windswept terraces on Santa Cruz and Santa Rosa often support a stress-tolerant community with only low-growing chamise and island scrub oak as major components. Santa Cruz Island oak *(Q. parvula)*, not pictured here, is another shrubby species restricted to Santa Cruz Island and a few coastal areas of Santa Barbara County. This oak is closely related to interior live oak *(Q. wislizenii)* on the mainland. Like

Plate 286. Island ceanothus (*Ceanothus arboreus,* Rhamnaceae), February to May.

Plate 287. Island scrub oak (*Quercus pacifica*, Fagaceae), March to April.

a number of other island species, Santa Cruz Island oak differs in part from its mainland relative in having larger leaves.

Four species of manzanita are found only on single islands. Santa Cruz Island manzanita *(Arctostaphylos insularis)* and McMinn's manzanita *(A. viridissima)* are tall shrubs restricted to Santa Cruz Island. These species appear to be relatively faithful to volcanic and metamorphic parent materials. Santa Rosa Island manzanita *(A. confertiflora)* is a smaller shrub on sedimentary rock substrates and has an extremely limited range even on this single island. Santa Catalina Island manzanita *(A. catalinae)* is a tall shrub on volcanic substrates that approaches the form of a small tree up to 15 feet in height (pl. 288). None of these species has a basal burl or root crown.

Disturbed sites of island chaparral throughout the Channel Islands, as well as island woodlands, have scattered populations of island tree mallow *(Lavatera assurgentiflora)*. This soft-wooded shrub reaches up to 10 feet in height and displays broad-lobed leaves and masses of large white to pinkish red flowers with purple veins (pl. 289). Island tree mallow

Plate 288. Santa Catalina Island manzanita (*Arctostaphylos catalinae,* Ericaceae), February to March.

Plate 289. Island tree mallow (*Lavatera assurgentiflora,* Malvaceae), March to November.

takes two distinctive forms, a more arborescent form on the northern islands and a shrubbier form with larger and glossier leaves on the southern islands. Because of its graceful form and showy flowers, island tree mallow has been widely planted in Southern California.

Island Woodlands

Deep soils on north-facing slopes or within ravines and narrow canyons, particularly at higher elevations, support local stands of a dense woodland community with a canopy 15 to 45 feet in height (pl. 290). These island woodlands are best developed on Santa Cruz, Santa Rosa, and Santa Catalina, although small stands of these woodlands can be found on San Clemente. These woodlands, once more extensive, have been heavily impacted by human activities and grazing over the past two centuries. Many stands have been strongly impacted as well by erosion of surface soils resulting from the grazing pressure, which reduced the understory cover. Limited seedling success in these stands today threatens the stability of island woodland communities.

Plate 290. Mixed woodland of oaks and fern-leaved Catalina ironwood on Santa Cruz Island, north of Prisoner's Harbor (photograph by Laurel Woodley).

Plate 291. Live oak woodland on Santa Cruz Island (photograph by Mildred Mathias).

Several species of evergreen oaks can provide a tall canopy up to 60 feet high or more in island woodlands (pl. 291). These include the coast live oak *(Quercus agrifolia)* and the island oak *(Q. tomentella)* (pl. 292). The understories of these oak woodlands contain many of the same species encountered in similar habitats on the mainland.

Another canopy tree dominating scattered island woodlands is Catalina ironwood *(Lyonothamnus floribundus)*, which can reach 40 to 60 feet in height. The fossil record of this fascinating species shows a past widespread distribution on the mainland, but today it has a restricted range on just the four largest Channel Islands. The unusually shaped leaves of Catalina ironwood show an interesting pattern of variation between islands. The typical subspecies occurring on Santa Catalina has simple oblong leaves three to six inches in length. The leaves of the subspecies on Santa Cruz, Santa Rosa, and San Clemente have fernlike divisions into two levels of leaflets (pl. 293). The unusual foliage and attractive peeling red bark have made this tree a popular landscape plant in Southern California.

Plate 292. Island oak (*Quercus tomentella*, Fagaceae), April to May.

Plate 293. Fern-leaved Catalina iron-wood (*Lyonothamnus floribundus* subsp. *asplenifolius*, Rosaceae), May to June.

One of the most interesting near endemics in island woodlands is a small deciduous tree, Catalina crossosoma *(Crossosoma californicum)*; it is one of a handful of species that form a family unique to California and adjacent parts of the southwestern United States and Mexico (pl. 294). This species occurs on Santa Catalina and San Clemente and on Guadalupe

Plate 294. Catalina crossosoma (*Crossosoma californicum,* Crossosomataceae), February to May.

Island off the coast of Mexico, and there is a small population on the Palos Verdes Peninsula on the mainland.

Several smaller trees and arboreal shrubs reach up to 30 feet in height within stands of island woodland and occur on all of the major islands. Tall individuals of toyon and Catalina cherry are common. This lower canopy level also includes several endemic arboreal species. Island redberry *(Rhamnus pirifolia)* is a close relative of mainland redberry species (pl. 295). Like many of the island endemics, island redberry can take on an arboreal form, although it is frequently shrubby on Santa Cruz. Valley oak *(Quercus lobata)* can be found on Santa Cruz and Santa Catalina but occurs most typically as scattered trees in grasslands, oak woodlands, or pine woodlands, rather than in a distinctive community as on the mainland. It is interesting to speculate how a typically mainland oak with large acorns became established on just these two islands. MacDonald oak *(Q. × macdonaldii)* is a low-growing deciduous tree common in woodland communities on Santa Cruz, Santa Rosa, and Santa Catalina. Not a true species, MacDonald oak originated as a hybrid between island scrub oak and a deciduous tree oak, probably valley oak.

As described in chapter 6, two of the mainland species of coastal pines form small woodlands on the northern Channel Islands. Bishop pine *(Pinus muricata)* can be found on both Santa Cruz and Santa Rosa, and Torrey pine *(P. torreyana)* occurs on the latter island.

Plate 295. Island redberry (*Rhamnus pirifolia,* Rhamnaceae), March to April.

Riparian Woodlands

The relatively small size of island watersheds combined with low rainfall limits the development of riparian woodlands and associated wetlands. Only Santa Cruz, Santa Rosa, and Santa Catalina have good examples of riparian communities. The dominant species are familiar from mainland riparian communities and all have wind-dispersed seeds. These include Fremont cottonwood *(Populus fremontii)*, black cottonwood *(P. trichocarpa)*, and three willows—arroyo willow *(Salix lasiolepis)*, red willow *(S. laevigata)*, and sandbar willow *(S. exigua)*. Mexican elderberry *(Sambucus mexicana)* is also common in this community, whereas bigleaf maple *(Acer macrophyllum)* is restricted to a few sites on Santa Cruz. Also interesting is the absence from the islands of several common mainland riparian trees—California or western sycamore *(Platanus racemosa)*, white alder *(Alnus rhombifolia)*, California walnut *(Juglans californica)*, and California bay *(Umbellularia californica)*.

Concerns about Alien Species

Thousands of plant species native to other parts of the world have been deliberately introduced into gardens and agricultural areas of Southern California, and hundreds more have arrived as hitchhiking seeds. We call these alien or nonnative plant species because their natural range of occurrence does not include our geographic area and their presence is due to intentional or accidental human activities. Most of these species only survive because we cultivate, feed, water, or otherwise care for them.

Perhaps 10 percent of these nonnative species adapt well to our environment and are able to reproduce and sustain their populations over multiple life cycles in our gardens without our direct help. We call these naturalized plant species. Reproduction by naturalized species is often adjacent to parent plants and does not necessarily involve any colonization of new areas. Many of our flowering garden annuals and bulbs reseed or reproduce well in this way. Our concerns arise with invasive alien species that disperse seeds or propagules (vegetative tissue with the ability to grow roots) widely and produce offspring that successfully colonize areas at considerable distances from the parents. These offspring are often produced in large numbers and thus have the potential to spread over extensive areas. Experience has shown that about one in 10 naturalized species becomes invasive in that it becomes established in and around native plant communities.

It may be no surprise to learn that the great majority of our invasive plant species in Southern California have their origins in the Mediterranean region, one of the four other regions of the world with a climate like our own. It was in the Mediterranean Basin that humans first developed extensive agricultural systems that allowed the rise of urban civilizations thousands of years ago, and it was here that groups of plant species evolved to take advantage of the regular disturbance and abundant nutrients and water associated with agricultural activities. After arriving in California and other mediterranean climate regions of the world, these well-adapted "weeds" spread rapidly along with the arrival of modern agriculture and the disturbances caused by cattle and sheep grazing.

Most of the invasive alien plants in our region are weedy species strongly associated with disturbed habitats such as roadsides, trails, cleared areas, and waste places. Familiar examples of these are black mustard *(Brassica nigra)* (pl. 296), castor bean *(Ricinus communis)* (pl. 297), and poison hemlock *(Conium maculatum)* (pl. 298). Although troublesome in a number of respects, particularly in their unsightly growth and promotion of fire along roadsides, these species do not generally invade natural habitats. They are probably restricted to disturbed habitats because they require the extra water or nutrients often provided by runoff along roads and trails, and because they are unable to compete directly with native species in undisturbed areas. Nevertheless, there is potential for some of these species to become more invasive in

Plate 296. Black mustard (*Brassica nigra,* Brassicaceae), not native, April to July.

Plate 297. Castor bean (*Ricinus communis,* Euphorbiaceae), not native, all year.

Plate 298. Poison hemlock (*Conium maculatum,* Apiaceae), not native, May to July.

the future, particularly if seed dispersal is a more important factor in distribution than microhabitat.

Although Southern California has to date largely been spared the horror stories of invasions such as those of acacias in South Africa, kudzu *(Pueraria montana)* in the southeastern United States, and Russian knapweed *(Acroptilon repens)* in our Northwest, our natural environments have been dramatically and significantly impacted by invasive species. A number of aliens have aggressively invaded our natural plant communities and have brought or threaten to bring irreversible changes to these ecosystems. They can have profound ecological impacts by physically crowding out native species by their aggressive growth, altering natural fire frequencies or intensities, changing soil nutrient availability, or competing aggressively for limited soil moisture. Many of our communities, especially grasslands, riparian woodlands, and coastal habitats, have been significantly transformed by invasives.

Grassland Invasions

We have already discussed in chapters 6 and 7 how the under-stories of oak savannas and grasslands in California are dom-inated by a matrix of alien annual grasses and broad-leaved herbs native to the Mediterranean Basin. Although their in-troduction was unintentional, these annuals rapidly spread to replace the native perennials in our grasslands and funda-mentally transform their environment. Much of the problem of poor reproduction by valley and blue oaks *(Quercus lobata and Q. douglasii)* in California over the last century has been caused by the manner in which these alien grasses compete with oak seedlings for moisture in the upper levels of soil.

Yellow star thistle *(Centaurea solstitialis)*, a weedy member of the sunflower family (Asteraceae) armed with needle-sharp spines, has become an aggressive invader of California grasslands (pl. 299). With stems as tough as rawhide cord, star thistle tangles mowers and other farm equipment, and pre-vents cows and other grazing animals from feeding on associ-ated grasses. Beyond the severe economic damage it causes to rangelands, the biggest reason to hate this plant comes from

Plate 299. Yellow star thistle (*Centau-rea solstitialis*, Asteraceae), not native, May to October.

the experience of trying to walk through a field of it. The spines can painfully penetrate even thick pants and shoes, From its humble beginnings in central California in the middle of the nineteenth century, yellow star thistle has expanded to dominate more than eight million acres in the northern and central parts of our state. Now, in just the past few years, it has relentlessly advanced into Southern California.

Riparian Invasions

Early European settlers in California brought with them a large bamboolike grass called giant reed *(Arundo donax)* to stabilize eroding stream banks. Giant reed proved spectacularly successful at this role, forming immense thickets 20 to 30 feet in height that tenaciously held the soil. However, sections of stems and roots easily fragmented, were carried downstream by floods, and showed equal success in colonizing wherever they came to rest. The result has been the establishment and growth of dense stands of giant reed along many lowland rivers throughout California. In the basin of the Santa Ana River in Southern California, for example, more than 10,000 acres are dominated by giant reed. These massive stands exacerbate flood problems by choking stream channels, create fire hazards in stream habitats otherwise relatively free of flammable tissues, and destroy native riparian habitat for rare and endangered species of birds and other wildlife. Moreover, billions of gallons of water are lost in Southern California each year from transpiration by the masses of giant reed choking many of our rivers and streams.

Two other invasive species show every indication of rapidly increasing their influence along streams in Southern California. Sticky eupatorium *(Ageratina adenophora),* an escapee from cultivation, is a shrubby herbaceous perennial native to Mexico that forms dense thickets of growth four to six feet in height in the understories of riparian woodlands

Plate 300. Sticky eupatorium (*Ageratina adenophora*, Asteraceae), not native, all year.

Plate 301. German ivy (*Senecio mikanoides*, Asteraceae), not native, December to March.

(pl. 300). It is well established in scattered canyons of the Transverse Ranges and in the Hollywood Hills, where it chokes out virtually all other growth. Equally aggressive is German ivy *(Senecio mikanoides)*, a luxuriant herbaceous perennial vine from South Africa (pl. 301). Once established, German ivy rapidly grows to literally engulf other plants in its vicinity.

Chaparral Invasions

Undisturbed chaparral stands, perhaps because of the stressful growing conditions under which they occur, have been relatively resistant to invasions. There are indications, however, that Spanish broom *(Spartium junceum)* is beginning to become established widely (pl. 302). This woody shrub has

Plate 302. Spanish broom (*Spartium junceum,* Fabaceae), not native, April to June.

been widely planted along the shallow soils of road cuts because as a member of the legume family (Fabaceae) it has the ability to fix atmospheric nitrogen in its root nodules. Related woody species of legumes from southwestern Europe have proved to be aggressive invaders in northern California.

Coastal Invasions

Coastal dunes, bluffs, and beach terraces in Southern California have been heavily impacted by invasive alien species. The widespread establishment of a number of these species is described in chapter 2. The greatest concern has been with the iceplant, or hottentot-fig *(Carpobrotus edulis)*, which was widely planted on coastal dunes in past years to control erosion (see pl. 16). It has proved very poor at this task because of its shallow root system, while at the same time it has aggressively expanded its cover and choked out native species. There are now widespread programs to use tarping and herbicides to remove this species and encourage growth of native dune species.

Serious problems of expansion of invasive aliens have occurred in recent decades along coastal bluffs. The most apparent of these aliens are two large perennial grasses, pampas

Plate 303. Pampas grass (*Cortaderia jubata*, Poaceae), not native, all year.

Plate 304. Fountain grass (*Pennisetum setaceum*, Poaceae), not native, all year.

Plate 305. Fennel (*Foeniculum vulgare*, Apiaceae), not native, May to September.

grass (*Cortaderia* spp.) (pl. 303) and fountain grass *(Pennisetum setaceum)* (pl. 304), which have widely escaped from cultivation and colonized large expanses of the coastal bluffs of Southern California.

Beach terraces and coastal flats have not been spared. Fennel *(Foeniculum vulgare)* is the most serious alien invader (pl. 305). Although a widespread roadside weed throughout Southern California, fennel takes on an ugly disposition on the Channel Islands and scattered coastal sites such as the Palos Verdes Peninsula, where it forms massive impenetrable thickets five to six feet or more in height. Extensive programs are under way to learn how to control this growth. Also causing severe problems on areas of the Channel Islands is another of the South African iceplants, finger mesemb *(Malephora crocea)* (see pl. 26). Expanses of coastal bluffs on Anacapa Island are covered completely by dense masses of this low succulent species.

Managing Invasive Species

The invasion potential of alien species is not always easily predictable. Rapid maturity to reproductive age, production of large numbers of seeds or vegetative propagules, and effective seed dispersal are common but not universal attributes of invasive species. Moreover, many naturalized species do not expand their range for many years and then suddenly become invasive. These changes may be due to genetic developments in their populations or to the arrival of a critical pollinator, seed dispersal agent, or symbiont. Thus, caution is always important in making predictions about potential invaders.

Controlling invasive species requires focused but flexible management approaches. For existing invasive species, triage is widely used to separate three categories of problems. One group, exemplified by the alien annual grasses, is so entrenched that there is little likelihood of control. At the other extreme are naturalized species not yet showing any indications of invading natural communities or rapidly expanding their ranges. Regular monitoring of these species may be justified. The most significant group comprises those species that are beginning to invade aggressively but may be controlled by mechanical, chemical, or biological means. Giant reed, hottentot-fig, and German ivy are examples of invasive aliens for which active control programs have been initiated.

Much of the biodiversity of Southern California results from the varied topography, climate, and geology of our region. Remarkably, this biodiversity exists adjacent to the second-largest urban center in the nation. Yet as our urban core expands and suburban outlying areas are developed, enormous threats to biodiversity arise. Four of the 10 counties with the largest numbers of rare and endangered plant and animal species in the continental United States are in Southern California, including Los Angeles, Santa Barbara, San Diego, and San Bernardino Counties. Monterey and Sonoma Counties in California also make the top 10 list.

Rare and Endangered Species

Rarity is not as simple a concept as it might seem. Species can be rare in a variety of ways, and these can have an impact on conservation strategies. A species can be rare because of a highly specific habitat requirement, a highly restricted geographic range, very small local population size, or some combination of these factors. This matrix can lead to a variety of categories of rarity, with some more significant than others. Many species in Southern California, for example, are rare here because of specific habitat requirements or very small population sizes, but are more common in their range outside of our region. These species are generally of less concern than are rare species that occur only in Southern California.

Approximately one-fourth of our more than 2,000 Southern California plant species are rare, endangered, or highly restricted in distribution. We describe a few of these here as examples of the differing manners in which species can be rare or endangered.

The first group includes species that have fairly wide geographic ranges of occurrence across our region, as well as fairly large population numbers, but are restricted to highly specialized habitats. A good example is Braunton's milkvetch

(Astragalus brauntonii), a tall herbaceous perennial found across Ventura, Los Angeles, and Orange Counties (pl. 306). This species is almost entirely restricted to coastal sage scrub and chaparral communities often on substrates of limestone, a relatively uncommon parent rock in our region. The hard-coated seeds of Braunton's milkvetch survive for many years in the soil and are stimulated to germinate by the heat of a fire. In this way, small populations have been known to increase to hundreds of individuals in the first year after an area burns. Urban expansion, habitat fragmentation, and fire control practices all endanger the few populations of this species.

Another locally abundant species in a highly restricted habitat is Orcutt's brodiaea *(Brodiaea orcuttii),* a vernal pool endemic ranging from Orange and western Riverside Counties southward through San Diego County into northwestern

Plate 306. Braunton's milkvetch (*Astragalus brauntonii,* Liliaceae), April to July.

Plate 307. Orcutt's brodiaea (*Brodiaea orcuttii,* Liliaceae), April to July.

Baja California (pl. 307). This species uses its bulblike corm to survive dry summer conditions around the margins of vernal pools. Orcutt's brodiaea, once widespread and flowering in large masses in spring, has been greatly reduced in abundance in recent years by the loss of its vernal pool habitat to urbanization.

Although also relatively widespread, Lyon's pentachaeta (*Pentachaeta lyonii*) is extremely rare despite a lack of obvious restriction to a highly specialized habitat (pl. 308). This tall annual member of the sunflower family (Asteraceae) is known from just five small populations in the Santa Monica Mountains and the western Simi Hills, where it has been

Plate 308. Lyon's pentachaeta (*Pentachaeta lyonii,* Asteraceae), March to April.

found growing in small grassland pockets near chaparral. It was once collected on the Palos Verdes Peninsula and on Santa Catalina Island but appears to no longer occur at either site. Urban expansion into foothill areas is a serious threat to remaining populations, as are weedy invasives in their grassland habitat.

Another group consists of species that are rare or endangered because of very limited geographic distribution. In most but not all cases these species are restricted to a specific habitat of limited occurrence, where they may have large or very small population numbers. Verity's dudleya *(Dudleya verityi)* is a small succulent rosette plant whose few small populations lie in a narrow band a few miles long on Conejo Mountain in the western Santa Monica Mountains (pl. 309). Its occurrence is restricted to north-facing slopes on volcanic substrates. This same limited area is the only site of occurrence for the rare Conejo buckwheat *(Eriogonum crocatum)*, a small subshrub growing on these same volcanic outcrops (see the chapter frontispiece).

Otay manzanita *(Arctostaphylos otayensis)* presents a similar example. This shrub is known only from small populations occurring on nutrient-poor volcanic soils on and adjacent to Otay Mountain in San Diego County. Santa Susanna

Plate 309. Verity's dudleya (*Dudleya verityi*, Crassulaceae), May to June.

tarweed *(Hemizonia minthornii)* is a small shrub with resinous foliage known only from a few scattered localities in the Santa Susanna Mountains, Simi Hills, and Santa Monica Mountains (pl. 310). Where it does occur, however, usually on hard sandstone outcrops, this species can be locally abundant.

Plate 310. Santa Susanna tarweed (*Hemizonia minthornii,* Asteraceae), July to October.

Plate 311. Catalina Island mountain mahogany (*Cercocarpus traskiae,* Rosaceae), March.

Perhaps the rarest tree species in California, and arguably the most endangered plant species in its native habitat, is Catalina Island mountain mahogany *(Cercocarpus traskiae)* (pl. 311). Never common, this local endemic from Santa Catalina Island was pushed to the brink of extinction by feral goats and pigs. With ongoing reductions in the numbers of these grazing animals, there is some optimism that new seedlings will become established.

Conserving Natural Communities

There is both good news and bad news regarding biodiversity in Southern California. A relatively large portion of the land area of California as a whole enjoys protected status and thus enhanced preservation of biodiversity. Including multiple-use recreation areas, where preservation is not the only goal, 12 percent of the state falls within such parks and reserves. Although this statewide figure is excellent, it tells only part of the story. More than 80 percent of the area of high mountain conifer forests and alpine habitats is protected, but threats from urbanization and other human impacts are very small for these regions. Wetlands, riparian woodlands, coastal ecosystems, native perennial grasslands, and vernal pools all face a different reality.

Despite the keystone significance of wetlands, for example, less than five percent of the original area of these ecosystems in Southern California remains in natural condition today. Nearly 25 percent of the endangered plant and animal species in California live in these wetland habitats, where threats of development continue. The El Segundo dunes and associated wetlands, for example, once covered more than 36 square miles on the ocean margin of Los Angeles, but less than one percent of this habitat remains today. Less than two percent of the original extent of riparian woodlands and less than one percent of perennial grasslands in the state remain in natural condition, and yet few of these remaining areas are protected.

Rapid urban and suburban expansion along the Southern California coast and inland in western Riverside and San Diego Counties have led to a loss of the vast majority of coastal sage scrub. This loss and the presence of a number of endangered species have spurred efforts in recent years to protect more of these habitats. Much of this conservation effort has been directed at protecting two important but increasingly rare vertebrate species, the California Gnatcatcher *(Polioptila californica)* and Stephens' Kangaroo Rat *(Dipodomys stephensi),* but many rare plant species also occur only in this habitat.

The Future

Threats to biodiversity remain acute in Southern California, and more specifically the urban centers of the Los Angeles Basin, Orange County, and the greater San Diego area. These are some of the most rapidly expanding urban areas in the United States. A robust economy fed by broadly based aerospace, entertainment, biotechnology, electronics, and import-export industries offers a magnet for economic growth and immigration. A critical component of this urban expansion has been the multiple nodes of development and home construction that have produced increasing fragmentation of natural areas. This can be seen very well in the Santa Monica Mountains to the west of the Los Angeles Basin. Here, many small watersheds or habitat islands have become separated from other natural areas by surrounding developments. The natural areas that do remain in the region, therefore, are becoming increasingly subdivided into unconnected patches. Such small pockets of remaining habitats restrict migration and gene flow between remaining populations of plants and animals.

Continued habitat loss and fragmentation threaten the long-term existence of many native species and pose the greatest threats to biodiversity in this area. Many plant and animal species once common in Southern California have been extirpated from the state, although some survive elsewhere. These include charismatic species such as the California Grizzly Bear, our state animal, which was once common in the chaparral and oak woodlands of our region. Other large mammals, such as Mountain Lions, Bobcats, and Badgers, and species with less mobility, such as amphibians and some reptiles, are at acute risk today and may be vulnerable to extinction by chance demographic, environmental, and genetic events in fragmented areas. Moreover, fragmentation not only jeopardizes wildlife populations but also provides expanded points of entry for invasive nonnative plant and animal species.

Effective programs to preserve and enhance biodiversity in Southern California must rely not just on public agencies but on public/private partnerships as well. The prospects for real progress toward such goals are excellent. Popular interest in the environment is strong in California, and nowhere in the world is there greater activity by grassroots citizen groups and nongovernment organizations working for preservation. The strength of the California economy is a mixed blessing in this respect. The state represents the fifth-largest economy in the world today, and Los Angeles County alone is the twelfth largest. The resources generated by this level of economic activity and popular interest in the environment should allow the region to be at the forefront of worldwide conservation efforts. At the same time, economic growth and urban expansion involve strong and continuing threats to regional biodiversity.

WHERE TO EXPERIENCE SOUTHERN CALIFORNIA PLANT COMMUNITIES

The table on pages 286–289 lists public parks and preserves that offer good examples of characteristic plant communities. These range in size from small coastal reserves to large national forests. A dot in the column indicates that good examples of this community are present. Green distinguishes woodland communities; brown distinguishes wetlands.

woodlands wetlands	Coastal Habitats	Sage Scrub	Chaparral	Live Oak Woodland	Valley Oak Savanna
Santa Barbara County					
Los Padres National Forest			•	•	•
Refugio State Beach	•		•		
El Capitan State Beach	•				
Goleta Slough					
Carpenteria State Beach	•				
Carpenteria Salt Marsh	•				
Ventura County					
McGrath State Beach					
Mugu Lagoon					
Santa Monica Mountains NRA*	•	•	•	•	•
Point Mugu State Park	•	•	•	•	
Leo Carillo State Park		•			
Ahmanson Ranch		•	•		•
Los Angeles County					
Point Dume State Beach	•				
Malibu Creek State Park			•	•	•
Malibu Lagoon State Beach	•				
Topanga State Park		•	•	•	
Ballona Wetlands					
Angeles National Forest			•	•	
Chino Hills State Park				•	
Verdugo Hills			•		
Orange County					
Bolsa Chica State Beach					
Crystal Cove State Park	•	•	•		
Newport Backbay/San Joaquin Marsh					
Riverside County					
San Bernardino National Forest**		•	•	•	
Mount San Jacinto Park and Wilderness			•	•	

Blue Oak Woodland	Engelmann Oak Woodland	Walnut Woodland	Conifer Woodland	Riparian	Grassland	Salt Marsh	Freshwater Marsh	Vernal Pool
•			•	•				
				•			•	
				•				
						•	•	
				•				
						•		
				•			•	
						•		
		•		•	•	•	•	
				•	•			
				•	•			
				•				
						•		
				•				
						•	•	
			•	•				
•				•	•			
				•				
						•	•	
				•	•			
						•	•	
			•	•				
			•	•				

continued ➢

	Coastal Habitats	Sage Scrub	Chaparral	Live Oak Woodland	Valley Oak Savanna
Cleveland National Forest***			●	●	
Santa Rosa Plateau		●	●	●	
San Diego County					
Palomar Mountain State Park			●	●	
Cuyamaca Rancho State Park			●	●	
San Onofre State Beach	●				
Torrey Pines State Park and Beach	●	●			
Cabrillo National Monument	●	●			
Silver Strand State Beach	●				
Tijuana Slough NWR	●				

* Straddles Ventura and Los Angeles Counties.

** Straddles Santa Barbara and Riverside Counties.

*** Straddles Riverside, San Diego, and Orange Counties.

Blue Oak Woodland	Engelmann Oak Woodland	Walnut Woodland	Conifer Woodland	Riparian	Grassland	Salt Marsh	Freshwater Marsh	Vernal Pool
			•	•				
	•			•	•			•
			•	•				
			•	•				
						·		
			•					
						•		

REFERENCES AND FURTHER READING

Barbour, M. G., and J. Major, eds. 1977. *Terrestrial vegetation of California.* New York: John Wiley.

Beauchamp, R. M. 1986. *A flora of San Diego County, California.* San Diego: Sweetwater River Press.

Bossard, C. C., J. M. Randall, and M. C. Hoshovsky. 2000. *Invasive plants of California wildlands.* Berkeley and Los Angeles: University of California Press.

Dale, N. 1985. *Flowering plants of the Santa Monica Mountains.* Santa Barbara: Capra Press.

Dawson, E. Y., and M. S. Foster. 1982. *Seashore plants of California.* Berkeley and Los Angeles: University of California Press.

Hickman, J. C., ed. 1993. *The Jepson manual: Higher plants of California.* Berkeley and Los Angeles: University of California Press.

Lanner, R. M. 1999. *Conifers of California.* Los Olivos, Calif.: Cachuma Press.

McAuley, M. 1985. *Wildflowers of the Santa Monica Mountains.* Canoga Park, Calif.: Canyon Publishing.

Munz, P. A. 2003. *Introduction to shore wildflowers of California, Oregon, and Washington.* Berkeley and Los Angeles: University of California Press.

Orrnduff, R., P. M. Faber, and T. Keeler-Wolf. 2003. *Introduction to California plant life.* Berkeley and Los Angeles: University of California Press.

Pavlik, B. M., P. C. Muik, S. Johnson, and M. Popper. 1991. *Oaks of California.* Los Olivos, Calif.: Cachuma Press.

Raven, P. H., H. J. Thompson, and B. A. Prigge. 1986. *Flora of the Santa Monica Mountains.* 2nd ed. Special Publication No. 2. Los Angeles: Southern California Botanists.

Sawyer, J. O., and T. Keeler-Wolf. 1995. *A manual of California vegetation.* Sacramento: California Native Plant Society.

Schoenherr, A. A. 1992. *A natural history of California.* Berkeley and Los Angeles: University of California Press.

Schoenherr, A. A., C. R. Feldmeth, and M. J. Emerson. 1999. *Natural history of the islands of California.* Berkeley and Los Angeles: University of California Press.

Stuart, J. D., and J. O. Sawyer. 2001. *Trees and shrubs of California.* Berkeley and Los Angeles: University of California Press.

ADDITIONAL CAPTIONS

PAGES x–1 California poppies on Munz Ranch Road near the Poppy Reserve, Los Angeles County.

PAGES 10–11 Coastal sand dunes near Point Mugu Lagoon, Ventura County.

PAGES 26–27 Coastal sage scrub community at Point Mugu State Park, Ventura County.

PAGES 60–61 Mixed chaparral in the Santa Monica Mountains.

PAGES 108–109 Chaparral fire near the campus of Pepperdine University, Malibu, in 1985 (photograph by Stephen Davis).

PAGES 136–137 Mixed woodland of Engelmann oak and coast live oak on the Santa Rosa Plateau.

PAGES 168–169 Community of nonnative annual European grasses at Nicholas Flats in the Santa Monica Mountains.

PAGES 196–197 Riparian woodland with white alder, California sycamore, and California bay at Solstice Creek in the Santa Monica Mountains.

PAGES 222–223 Salt marsh community in the estuary at McGrath Beach State Park, Ventura County.

PAGES 244–245 Aerial view of Anacapa Island.

PAGES 262–263 *Arundo donax* (not native), also called giant reed, now widely established along streams and rivers throughout Southern California.

PAGES 274–275 Conejo buckwheat *(Eriogonum crocatum),* known only from restricted populations in the western Santa Monica Mountains of Ventura County.

INDEX

ABOUT THE AUTHORS

Philip W. Rundel is professor of biology in the Department of Ecology and Evolutionary Biology at the University of California, Los Angeles. He is also director of the Stunt Ranch Santa Monica Mountains Reserve and senior investigator in the Center for Embedded Network Sensing at UCLA.

Robert Gustafson is the retired collections manager of the Botany Department at the Natural History Museum of Los Angeles County. He has been studying and photographing California flora for over 30 years. His photographs illustrate *Plants and Flowers of Hawaii* (1987).

Series Design:	Barbara Jellow
Design Enhancements:	Beth Hansen
Design Development:	Jane Tenenbaum
Composition:	Jane Rundell
Text:	9.5/12 Minion
Display:	ITC Franklin Gothic Book and Demi
Printer and binder:	Everbest Printing Company